Water Resources Management Sustaining Socio-Economic Welfare

GLOBAL ISSUES IN WATER POLICY

VOLUME 7

For further volumes:
http://www.springer.com/series/8877

Phoebe Koundouri • Nikos A. Papandreou
Editors

Water Resources Management Sustaining Socio-Economic Welfare

The Implementation of the European Water Framework Directive in Asopos River Basin in Greece

Springer

Editors
Phoebe Koundouri
Department of International and European Economic Studies (DIEES)
Athens University of Economics and Business
Athens, Greece

Grantham Research Institute on Climate Change and the Environment
London School of Economics and Political Science
London, UK

Nikos A. Papandreou
Andreas G. Papandreou Foundation
Athens, Greece

ISSN 2211-0631 ISSN 2211-0658 (electronic)
ISBN 978-94-007-7635-7 ISBN 978-94-007-7636-4 (eBook)
DOI 10.1007/978-94-007-7636-4
Springer Dordrecht Heidelberg New York London

Library of Congress Control Number: 2013954372

Printed on acid-free paper

Springer is part of Springer Science+Business Media (www.springer.com)

As always, this book is dedicated to Nikitas, my resilience, and Chrysilia and Billie, my happiness. Billie was born while I was working on the "Asopos" Project and this book.

Prof. Dr. Phoebe Koundouri

Preface

The overreaching aim of the book is to offer a river-basin management plan which is directly implementable and consistent with the European Union Water Framework Directive (EU WFD). This management plan consists of explicit technical and investment solutions, socio-economic and legal instruments, as well as institutional restructuring recommendations. Given the well-recognized challenge (both academically and empirically) of managing natural resources in a way that maximizes and sustains social welfare, this book is an invaluable point of reference to applied researchers and policy makers in the field of water resources management.

The first chapter is an introductory chapter that describes the water situation in Greece and assesses the potential of the timely implementation of the EU WFD. Special emphasis is given to the cost-recovery principle, and results of a quick appraisal are presented.

Chapter 2 introduces the case study area putting the emphasis on the particular pressures and impacts as well as the environmental functions and values of Asopos River and Oropos Lagoon. The next chapter (Chap. 3) presents the main water uses and pricing for water supply as far as industry and agriculture are concerned, while Chap. 4 completes the baseline information presenting the residential water use. Chapters 2, 3, and 4 offer the background information to be used in the following chapters.

Next, the emphasis is on the agricultural sector. Chapter 5 aims to estimate the farmers' valuation of the groundwater's shadow price for the region of Asopos. The objective of the microeconometric analysis is to uncover the patterns of groundwater use and farm efficiency. The chapter offers policy recommendations based on the principle of socio-economic sustainability and presents an estimation of groundwater for irrigation shadow price and how this can be used in the design of pumping taxes to reduce pollution and to increase farm efficiency.

Chapter 6 presents the results of an environmental valuation study (Choice Experiment) aimed to value environmental improvements in the Asopos water catchment to be used for policy purposes. The same holds for Chap. 7 which employs another method, that of Benefits Transfer, in order to estimate the environmental cost of damage due to the industrial activity. Results of both chapters are compared with interventions for improving the current situation.

The net cost suffered by the local farmers due to the environmental degradation is presented in Chap. 8 where policy recommendations are also offered. Finally, Chap. 9 presents an integrated water resources management plan in the Asopos river basin, while Chap. 10 offers recommendations regarding institutional changes and presents lessons learned for the rest of the Greek river basins.

Athens, Greece Phoebe Koundouri

Acknowledgments

The assistant editor of this book is Dr. Mavra Stithou. Without her excellent editorial work and dedication, this book would not have been completed.

Phoebe Koundouri

Contents

Contributors

Yannis Anastasiou holds an M.Sc. in Risk Analysis and Management from the Southampton Management School, University of Southampton (UK), and a B.A. in Economics from the University of Cyprus. He is currently an associate researcher with RESEES (Research tEam on Socio-Economic and Environmental Sustainability) at the Athens University of Economics and Business. He has published a number of research papers and book chapters and has participated in many research projects funded by the European Commission.

Antonis Antypas is a Researcher and Teaching Associate at the University of Piraeus. He holds a Ph.D. in Financial Econometrics and Software Development and an M.Sc. in Banking and Financial Management both from the University of Piraeus. He obtained his B.Sc. in International and European Economic Studies from the Athens University of Economics and Business. His research interests include time-varying parameter models, asset allocation under varying second moments, and Monte Carlo simulations – programming.

Osiel González Dávila is Research Associate at the Athens University of Economics and Business. His principal research interests lie in the fields of economics, environment, and development. He holds an M.Res. in Environment and Development from the University of Lancaster and an M.Sc. in Economics from the University of Essex and is a Ph.D. candidate of Economics at the University of London where he is also a Senior Teaching Fellow.

Charalampos Doulgeris received a B.Sc. in Agriculture, an M.Sc. in Agricultural Engineering and Water Resources, and a Ph.D. in Groundwater Hydrology from Aristotle University of Thessaloniki. He has practiced as a professional agricultural engineer in the area of water resources and environmental management since 2000 and is registered as a member of the Geotechnical Chamber of Greece and the Hellenic Hydrotechnical Association. His research experience is related to watershed modeling for water resources management and model development for the simulation of subsurface flow and contaminant transport.

Andreas Drichoutis is an Agricultural Economist. He is currently a Lecturer of Applied Microeconometrics at the Department of Economics at the University of Ioannina in Greece. He has been working for several EU or national projects and as a consultant for the Food and Agriculture Organization of the United Nations. His research interests are focused on the economics of nutrition/obesity, on contingent valuation and experimental auctions methods, on choice under risk, and on intertemporal decision making. He has published several peer-reviewed papers in peer-reviewed journals, and his work has been cited at least 150 times in scientific journals.

Sotiria Katsavouni received her B.Sc. in Environmental Studies from the University of Aegean in 1994. In 2003, she obtained her M.Sc. degree in Ecology and Sustainable Management of Ecosystems from Aristotle University of Thessaloniki. Since 1995, she works as a research scientist at the Greek Biotope/Wetland Centre in soil and water resources management. She has participated in research projects concerning wetlands restoration and rehabilitation, wetlands functions evaluation and creation wetland for wastewater treatment using constructed wetlands.

Phoebe Koundouri holds a Ph.D. (2000), M.Sc. (1996), and M.Phil. (1995) in Economics from the University of Cambridge (UK). She is Associate Professor at the Department of International and European Economic Studies, Athens University of Economics and Business (Greece) and the Director of the Research tEam on Socio-Economic and Environmental Sustainability (RESEES, http://www.aueb.gr/users/koundouri/resees/). She is also Senior Research Fellow at the London School of Economics and Political Science, Grantham Research Institute on Climate Change and the Environment, UK. Before, she held academic positions at the University of Cambridge, University College London, University of Reading, London Business School, and visiting positions at the Toulouse School of Economics, University of Cyprus, Cyprus Institute, and Open University of Cyprus. Her research focuses on Environmental and Natural Resources Economics and Econometrics. She has published 12 edited books and more than 250 articles in edited volumes and top-ranked academic journals. She is ranked among top 5 % authors in IDEAS (http://ideas.repec.org/e/c/pko155.html) and is a highly cited author in her scientific field (Harzing's Publish or Perish: more than 1800; Google Scholar: more than 1900; General Google: more than 43,900 hits). She is co-editor of the official journal of the European Association of Environmental and Resource Economists, associate editor and editorial board member in 13 academic journals in the field of economics, and has supervised a number of Ph.D. students, all of them now working in academia. In 2012, Phoebe Koundouri was elected Council Member of the European Association of Environmental and Resource Economists (EAERE, www.eaere.org) and she is currently the EAERE Newsletter Editor. She is also Core Member of the Scientific Committee of the John S. Latsis Public Benefit Foundation. Since 1996, Phoebe Koundouri has coordinated and/or participated in more than 50 research and consultancy projects on natural resources and environmental management funded by the European

Commission, World Bank, OECD, UN, NATO, WHO, as well as many other international and national organizations and governments of developed and developing countries. Over the last two decades, she has given academic and public lecturers all over the world, organized a number of international conferences, mainly in the field of water resource economics and management, and received scholarships and prizes for academic excellence from various academic institutions and foundations. She is included in various Who's Who and has recently (in 2012) received the 2009 Publication Award of the European Association of Agricultural Economists.

Yiannis Kountouris is a Lecturer in Environmental Economics at the Centre for Environmental Policy, Imperial College London. His research interests are the valuation of environmental resources and ecosystem services, the role of environment in policy making, and the role of culture in environmental preferences.

Theodoros Mavrogiorgis is an expert in highway design, traffic management, advanced transport telematics, transport planning and modeling, and environmental assessment. He has been project leader in various EU transport and energy related demonstration projects. He has conducted numerous environmental studies and environmental impact assessment studies. He holds an M.Sc. in Financial Analysis from the University of Piraeus, an M.Sc. (Eng.) in Highways and Transport from the City University of London, and a Diploma in Civil Engineering from the National Technical University of Athens.

Philippos Melissourgos has a B.Sc. in Economics from the Athens University of Economics and Business. His dissertation was supervised by Prof. Koundouri (during the academic year 2011–2012) and focused on the analysis of Water Resources Management in the Asopos River Basin. Philippos participated in a number of RESEES environmental economics and management related projects, coordinated by Prof. Koundouri. His contribution mainly focused on data collection and descriptive statistical analysis.

Aris Mousoulides is a Ph.D. candidate in Quantitative Tourism Analysis at the University of Middlesex (UK) and holds an M.Sc. in Hotel Administration from the University of Strathclyde (UK) and a Diploma in Hotel Management from the Les Roches Swiss Hotel Association Hotel Management School (Switzerland). He is also a Senior Lecturer and Campus Program Coordinator at Intercollege (Cyprus) and an associate researcher with RESEES (Research tEam on Socio-Economic and Environmental Sustainability) at the Athens University of Economics and Business. Formerly, he was Deputy Director General for the Association of Cyprus Tourist Enterprises (ACTE). He has published various papers and book chapters and participated in many projects funded by the European Commission.

Marianna Mousoulidou is Bank Manager at Laiki Popular Bank (Cyprus) and consultant, with more than 20 years of experience, and specializes in economics and finance. She holds an M.B.A. and B.A. in Economics from the University of Leicester (UK) a B.Sc. in Financial Services from the University of Manchester

(UK) and two professional degrees, one from the UK Chartered Institute of Bankers (ACIB) and a second from the Cyprus Securities and Exchange Commission. She has coauthored a number of academic papers and book chapters.

Dimitris Papadimos received his B.Sc. in Agricultural Engineering from the Department of Hydraulics, Soil Science and Agricultural Engineering, School of Agriculture, Aristotle University. In 1995, he obtained his M.Sc. from the same department. He worked for 3 years in Aristotle University of Thessaloniki on research involving rain-runoff relations and irrigation systems. Since 1997, he is working for Greek Biotope/Wetland Centre, Goulandris Natural History Museum (EKBY) as a research scientist on soil and water resources management. His recent work includes project management and application of hydrological models for wetland restoration.

Nikos A. Papandreou studied economics and political science at Yale and earned a Ph.D. in Economics from Princeton in 1986. After Greek military service and a few years at the World Bank, he received an M.F.A. in Creative Writing from Vemont University (1994) and turned to writing on a more full-time basis. He has written many books, most in Greek, while his essays, articles and short stories have appeared in prestigious anthologies and journals. He is a regular columnist for a number of Greek magazines such as *The Economic Review*. He is presently heading a project to save the Asopos River in the province of Viotia. He belongs to the board of directors of the Andreas G. Papandreou Foundation.

Kyriaki Remoundou is Lecturer in Economics at the School of Management and Business in Aberystwyth University, Wales. She holds a Ph.D. in Economics, an M.Sc. in International Economics and Finance, and a B.Sc. in International Economics and Finance from the Athens University of Economics and Business. She is interested in the fields of environmental economics, non-market valuation, experimental auctions, life satisfaction, and applied microeconometrics.

Riccardo Scarpa is a Ph.D. graduate from the University of Wisconsin-Madison (1999). He has held academic positions in Italy, the UK, Australia, and New Zealand. He coauthored more than 70 peer-reviewed research articles, with over 3,000 citations in Google Scholar, and currently has a Hirsch index of over 30. He has acted as a peer-reviewer for more than 70 academic journals and has extensive experience as a research project evaluator for various granting agencies. Ric supervised the Ph.D. research of more than 15 students and mostly works in the area of non-market valuation, especially focusing on environmental goods.

Ioannis Souliotis was born in 1990 in Trikala, Greece. In 2011, he obtained his B.Sc. in International and European Economic Studies from Athens University of Economics and Business, Greece. In 2012, he started his master's studies in Environmental and Natural Resource Economics at the University of Copenhagen, Denmark. His main research focus is related to the evaluation of non-market goods using Revealed Preferences techniques.

Mavra Stithou is an Environmental Economist. She holds a Ph.D. from the University of Stirling and an M.Sc. from the University of York. She specializes in water and marine policy, integrated socio-economic assessment of marine planning, the application of environmental valuation techniques (non-market valuation methods), cost-benefit analysis, and the use of benefits transfer.

Katerina Vasiliou is a Ph.D. candidate at the Laboratory of Geography and Spatial Analysis, Department of Geography and Regional Planning, National Technical University of Athens, and an associate researcher with RESEES (Research tEam on Socio-Economic and Environmental Sustainability) at the Athens University of Economics and Business. She has an M.Sc. in Urban Planning from the School of Architecture of the same university and a Diploma in Surveying and Rural Engineering from the School of Surveying and Rural Engineering from the same university. She is an experienced consultant and software analyst and has participated in many projects concerning environmental impact assessment of public projects.

List of Figures

List of Tables

Chapter 1
A Bird's Eye View of the Greek Water Situation: The Potential for the Implementation of the EU WFD

P. Koundouri, N. Papandreou, K. Remoundou, and Y. Kountouris

Abstract The Water Framework Directive (WFD) was formulated for addressing the weaknesses of the previous water-related directives. The main steps that WFD involves could be summarized in the setting of ecological standards, the identification of anthropogenic pressures and the adoption of corrective measures. This introductory chapter describes the water situation in Greece and assesses the potential of the timely implementation of the European Union's (EU) WFD. In this context, the significance of Asopos River Basin (RB) is put into perspective. More analytically, the chapter presents: (a) the employed methodology that enables rapid assessment of the *status quo* of the water situation in each Greek catchment, as compared to the requirements and targets of the EU WFD, (b) the implementation of this methodology on each of the 14 Greek River Basin Districts (RBDs) and (c) relevant empirical results. The main objective of the chapter is to present the rapid-appraisal methodology that was developed for the estimation of the cost-recovery level for water services in the 14 Greek RBDs. Results from this 'quick appraisal' clearly highlight the need for reforms in the current pricing policy and preparation of a package of measures,

P. Koundouri (✉)
Athens University of Economics and Business, Athens, Greece
e-mail: pkoundouri@aueb.gr

N. Papandreou
Andreas G. Papandreou Foundation, Athens, Greece
e-mail: tulayev@gmail.com

K. Remoundou
Aberystwyth University, Aberystwyth, UK
e-mail: kyriakiremoundou@yahoo.gr

Y. Kountouris
Imperial College London, London, UK
e-mail: i.kountouris@imperial.ac.uk

P. Koundouri and N.A. Papandreou (eds.), *Water Resources Management Sustaining Socio-Economic Welfare*, Global Issues in Water Policy 7,
DOI 10.1007/978-94-007-7636-4_1,

as proposed in Chaps. 9 and 10, in order for the water bodies to reach good ecological status and the water management to ensure full recovery of the cost of water services as required under article 11 of the WFD.

Keywords EU WFD • Asopos River Basin • Water quality • Cost recovery • Water pricing

1 Introduction

1.1 General Overview of the WFD

It is internationally recognized that water resources are necessary inputs to production in economic sectors such as agriculture (arable and non-arable land, aquaculture, commercial fishing, and forestry), industry (power generation) and tourism, as well as to household consumption (UNEP 2005).

Policy makers at European level have recognized the need to approach human activity and water resources in an integrated manner to achieve sustainable water resources management as laid down in the recently adopted EU-Water Framework Directive (WFD) (CEC 2000). The WFD (2000/60/EC) was adopted in October 2000, and it establishes a framework for European Community action in the field of water policy. The importance of water is crystallized in the first recital of the Directive. It states that "[W]ater is not a commercial product like any other but, rather, a heritage which must be protected, defended and treated as such" (CEC 2000, p. I.327/1).

The aim of the WFD is to establish a framework for the protection of inland surface waters, transitional waters, coastal waters and ground waters. Whereas previously adopted water-related EU Directives addressed individual issues, the WFD aims to provide an integrated framework for water resources management, both in terms of quality and quantity, to achieve the objective of good water status for all EU waters. Figure 1.1 presents the integrated character of WFD.

Hence, an examination of water policy through previous water directives, including the Nitrates Directive Nitrates Directive (91/676/EEC) and the Bathing Water Quality Directive Bathing Water Directive (2006/7/EC), demonstrates how current policy evolved from an emphasis on public health protection to environmental protection and finally, as formed today, to the notions of 'sustainable use' of water and an integrated ecosystem-based approach to water management. What is achieved from these changes is that although in the past EU legislation on water was focused on specific environmental problems related to water quality as far as for example drinking, bathing or freshwater fishing activities are concerned, emphasis is now placed on the improvement of the ecological quality of water and its eco-system functions, by using a broader and integrated approach involving both environmental quality objectives coupled with emission limit values.

The Directive calls for integrated catchment management plans to be prepared for all river basins in order to achieve Good Ecological Status (GES) in all EU

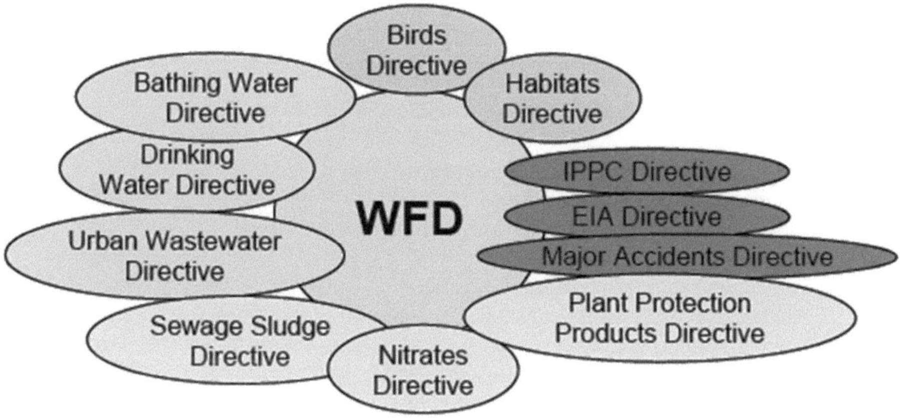

Fig. 1.1 WFD, a truly integrated Directive

waters by 2015. Particularly, according to Article 2 (18), '[G]ood surface water status' refers to the status achieved by a surface water body when both its ecological status and its chemical status are at least 'good'. As such, the Directive aims at achieving a minimum standard of 'good' and 'non-deteriorating' status, and sets common approaches and goals for water management in the EU Member State countries (MS) adopting a broader measure of water quality.

The suggested means to achieve that goal is the planning at the natural hydrologic (river basin) level/unit instead of other administrative or political boundaries and the implementation of pollution-control measures in cases where existing legislation on water quality and pollution is proved inadequate. Hence, an important change in water management policy is that the measures to achieve WFD objectives will be co-ordinated at the level of River Basin District (RBD) that will correspond to large catchment basins incorporating the smaller sub basins. In the case that a basin crosses national boundaries, the responsibility should be shared between governments and one single vision should be created.

For the assessment of quality, three main characteristics are considered. The first is that of biological quality elements. The parameters to be measured for river, lake and transitional waters are composition and abundance of aquatic flora (macrophytes) and benthic fauna (invertebrates) as well as the composition, abundance and age of structure of fish. In the case of the marine environment, instead of the 'fish' parameter the composition, abundance and biomass of phytoplankton is considered. The other two quality characteristics refer to elements that support biological elements. One is the physico-chemical elements such as condition of thermal, oxygen, salinity, acid, nutrient and transparency, and the other is hydromorphological elements that can include in the case of a river for example, the quantity and dynamics of water flow, its continuity, depth and width variation, and structure of the riparian zone. The Directive's goal is diversified in the case of 'artificial/modified' waters serving economic activities where the GES turns to 'good ecological potential' and in the

Table 1.1 WFD timetable

Year	Issue	Reference
2000	Directive entered into force	Art. 25
2003	Transposition in national legislation identification of RBDs and authorities	Art. 23 Art. 3
2004	Characterization of river basin: pressures, impacts and economic analysis	Art. 5
2006	Establishment of monitoring network Start public consultation (at the latest)	Art. 8 Art. 14
2008	Present draft river basin management plan	Art. 13
2009	Finalize river basin management plan including program of measures	Art. 13 & 11
2010	Introduce pricing policies	Art. 9
2012	Make operational programs of measures	Art. 11
2015	Meet environmental objectives	Art. 4
2021	First management cycle ends	Art. 4 & 13
2027	Second management cycle ends, final deadline for meeting objectives	Art. 4 & 13

Source: http://ec.europa.eu/environment/water/water-framework/info/timetable_en.htm

case of 'protected zones' (i.e., areas designed for the protection of habitats or species) and nutrient sensitive areas where more stringent requirements may be applied.

For its implementation, the Directive calls for the authority of each RBD to prepare and put into action a 6 year River Basin Management Plan that will include a description of the district's characteristics, the identification of protected areas, the impact and pressures of human activity on water status (point source and diffuse pollution, abstraction and land-use patterns), an economic analysis of the cost of the water, an estimation of the effects of existing legislation to achieve the objectives, and information on measures taken to achieve goals. In implementing the measures, MS are asked to take into account the principle of full recovery of costs of water services that will provide incentives for the efficient use of water by different users. At this stage, according to Article 14, public participation of all interested parties should contribute to the identification of measures to be adopted.

It should be noted that in this context, monitoring is central to the Directive and Article 8 includes several monitoring requirements, not only to determine the classification of waters' status but also to continue assessing the necessity for additional measures or ensure that mitigation measures are implemented. Thus, the main steps that the WFD involves could be summarized in the setting of ecological standards, the identification of anthropogenic pressures, and the adoption of corrective measures. Furthermore, the main change and innovation that the Directive brings is that it institutionalizes the ecosystem objectives and has, to some extent, a binding character. Hence, "for the first time in the EU Environmental Policy a legal text proposes economic principles and measures as basic instruments for the achievement of specific environmental objectives" (MoEPPW 2006, p. 233). For each MS there is a Common Implementation Strategy (CIS) and timetable as summarized in the following table (Table 1.1).

2 The Socio-Economic Aspects of the EU WFD

From an economic perspective, water resources are not efficiently allocated and may be overexploited due, to some degree, to the existence of market and government failures at different levels (local, national, international). This phenomenon primarily occurs because of the public good nature of water resources and secondly because of the complexity that characterizes water value (including use and non-use values), that does not allow it to be traded in markets as private goods. Brouwer et al. (2009, p. 13) argue that the main problem when considering economic choices related to water is that a competitive, freely functioning market does not exist for many water related uses because "water is an essential commodity such that the value for a basic survival amount is infinite; water has natural monopoly characteristics; property rights for water resources are often absent and difficult to define; water is a 'bulky' commodity, thereby restricting the development of markets beyond the local area".

As economic efficiency occurs at the point where net social benefits of an economic activity are maximized, or equivalently, when the marginal benefits are equal to marginal costs, in order to implement the most efficient social and economic policies that prevent the excessive degradation and depletion of environmental resources it is necessary to establish their full value, and to incorporate this into private and public decision-making processes (Birol et al. 2006). The WFD is one of the first European Directives to recognize explicitly the role of economics in reaching environmental and ecological objectives and it aims to correct for 'market or government failures' by managing water resources in a sustainable manner with the application of economic principles, approaches, tools and instruments at RBD level.

In particular, the EU WFD is one of the policy initiatives that aims to ensure the sustainable management and conservation of this valuable resource, along with other international efforts such as the 1971 Ramsar Convention on Wetlands of International Importance (Ramsar 1996). In order to achieve this, the WFD promotes the concept of water as an economic commodity, while maintaining its focus on its broader and often intangible value. However, given the different characteristics of demand for different uses of this resource related to location, quality, quantity and timing, any consideration of water as an economic good needs to ensure its commensurability in terms of a common denominator of place, form and time (Brouwer et al. 2009).

The procedure for the implementation of the economic analysis includes three steps (MoEPPW 2006, p. 233):

(i) Assessment of the current level of full cost recovery based on the economic analysis of water uses and long term forecasts of water supply and demand in each river basin district. The analysis aims at the development of the Baseline Scenario of the evolution of basic parameters affecting water demand and supply and necessary investments (Fig. 1.2).

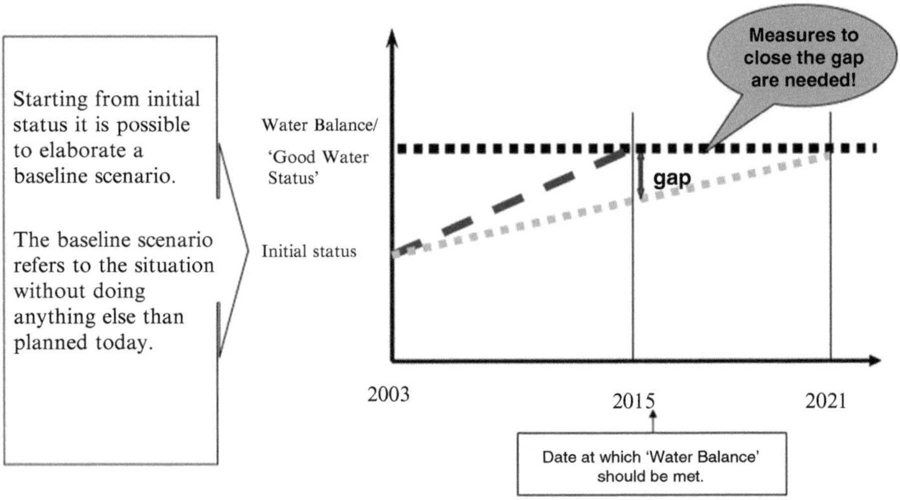

Fig. 1.2 Diagrammatical implementation of WFD (Source: http://www.aueb.gr/users/resees/uploads/wfd.pdf)

(ii) The Baseline Scenario is used for the assessment of the anticipated impacts on the quality of water bodies. Potential gaps in relation to the environmental objectives should be identified.
(iii) Assessment of the economic impacts from the application of program of measures.

Economic issues are mainly dealt with in Article 5 (Characteristics of the river basin district, review of environmental impact of human activity and economic analysis of water use), Article 9 (Recovery of costs for water services) Article 11 (Program of measures) of the Directive and Annex III (Economic analysis). These economic issues can be implemented (as indicated in the WATECO document, European Communities (2002)) by the use of the following river-basin specific three-step approach: (1) the economic characterization of water in the RBD, (2) the assessment of the recovery of the costs of water services, and (3) the economic assessment of potential measures for balancing water demand and supply. The first step involves (i) the estimation of the socio-economic significance of water uses using data concerning not only the water consumption, but also the production of pollution loads by the various activities and (ii) the investigation of the dynamics of key economic drivers that influence pressures and thus water status. As a result, the economic analysis must include the long term forecast of supply and demand, and estimates of volume, price and cost associated to water services where data is not available. An overview of the dynamics of the river basin should be provided, based on a top-down approach, forecasting changes in pressures based on the changes in key climatic and socio-economic drivers (e.g., population trends, trends for the major economic activities, land use changes, technological changes) in order to construct a baseline scenario.

The current level of recovery of costs of water services should then be assessed in a second stage to be used as a tool for appraising economic efficiency and equity and with a final aim of identification of least-cost measures to achieve sustainable water resources management in the final step. Therefore, the economic analysis reports should contain sufficient information on the significant drivers and pressures in each RBD and on the contribution of water uses in the recovery of costs in accordance with the polluter pays principle, to enable the selection of the program of measures on a cost effectiveness basis in 2010 (Annex III). Overall, economic principles are to be applied in four main areas within a river basin context (Morris 2004, p. 4):

- The estimation of the demand for water and the valuation of water in its alternative uses (Article 5)
- The identification and recovery of costs, environmental and resource, associated with water services, having regard for the polluter pays principle and the efficient use of water (Article 9)
- The use of economic appraisal methods to guide water resource management decisions (Article 11)
- The use of economic instruments to achieve the objectives of the WFD, including the use of incentive pricing and market mechanisms (Article 11)

Specifically Article 5 introduces the principle of economic analysis in water management and the assessment of the most cost-effective combination of measures in respect of water uses to be included in the program of measures under Article 11, based on estimates of the potential costs of such measures. Furthermore, the economic analysis is also expected to provide room for derogations under the umbrella of disproportionate costs. In relation to the latter concept, Article 4 states that exemptions are possible if the cost of reaching the GES is disproportionate.[1] However, in order to evaluate the extent to which this is the case and to assess 'disproportionality', one also has to know the costs and benefits associated with reaching environmental objectives, in both qualitative and quantitative terms. In order to pass the test, costs should exceed benefits by a significant margin in a cost-benefit framework.

Article 9 stresses the need for users (that is industries, farmers, and households) to be charged a price that reflects the full cost of the water services they receive. Full cost pricing is a mandatory part of the river management plan and according to the Directive's timetable MS should have introduced water pricing policies by 2010. According also to Article 9.1 environmental and resource costs must be taken into account for water services, according to provisions of Annex III and particularly to the polluter pay-principle.

Another clear aim of the economic analysis is to provide a preliminary selection of appropriate instruments and measures. This selection should consider the significant water management issues, pressures and impacts identified, and the measures and investments foreseen or already implemented (MoEPPW 2006). "Each measure should be assessed in terms of effect, cost and scale of application, in order to provide the basis for a more detailed assessment of costs and effects and ultimately

[1] Costs are considered as disproportionate if they exceed the monetized benefits of achieving 'good status' in a water body.

for the selection of appropriate supplementary measures to be included in the River Basin Water Management Plan" (MoEPPW 2006, p. 235). As a result, considering that each measure and water policy is associated with each own costs and benefits and should be judged in a long-run cost-effectiveness analysis basis to identify the optimal set of measures adequate to creative incentives for sustainable water resources use (Koundouri and Remoundou 2009; Birol and Koundouri 2008a; Koundouri 2007).

Hence, the tool kit of economic analysis includes the estimation of both direct and indirect costs and benefits to be considered in each management plan (Hanley and Black 2006). Regarding the nature of benefits, economic analysis will consider direct benefits such as reductions in the cost of drinking water treatment downstream when less pollution is discharged into a river and indirect benefits such as an increase in jobs if cleaner coastal waters lead to higher tourism levels. Furthermore, more difficult to quantify benefits, such as recreation and availability of healthy ecosystems, will also be included. It is regarded that the contribution of valuation methods can be useful in that respect. In general, this is an important but difficult task for river basin authorities, and it will involve them having to consider and evaluate costs and benefits - including environmental criteria. Hence, the concept of environmental and resource costs and benefits plays an important role in the economic analysis of the Directive and practical guidelines for their assessment have been developed (European Communities 2002; Brouwer et al. 2009). Furthermore, for supporting the coherent and harmonious implementation of the WFD, the MS developed the CIS (2004).

Overall, economics and their subset of environmental economics are expected to play an important and supportive role in WFD implementation (through Articles 9, 11 and 4), and in particular in justifying spending on environmental protection where applicable. Particularly focusing on the contribution of the valuation of benefits, it is regarded that their inclusion will facilitate water-related decision-making in different ways. However, according to the Commission's compliance report COM (2007 128 final), the economic assessment is one of the main shortcomings in the WFD implementation. In particular, although all MS have submitted country reports on Article 5 of WFD, half of them have not supplied information at all on cost recovery. This highlights the difficulties (informational and methodological) that MS face in implementing the economic aspects of the Directive. These difficulties will be exacerbated by the requirements of Article 9 of the Directive, which indicates that by 2010 MS should introduce pricing policies and economic instruments with the element of cost-recovery for the benefit of the environment.

3 The Need for a 'Quick Appraisal'

A rapid-appraisal approach was dictated by severe information deficiencies and limited time frame. At the time of the current study (November 2007) the European Commission had already initiated legal action of 'Non-Conformity' with the requirements under Article 5 against Greece (case A2005/2317). The time frame for this study was thus defined by the Ministry of Environment, Physical Planning and Public Works[2] (MoEPPW), who financed and supervised the study, to 2 months. Meanwhile the significantly low level of available information posed a further constraint in the study. Preliminary analysis of water uses, pressures and impacts, under the first step of the implementation procedure, aiming to inform and guide the subsequent economic analysis was piecemeal. The only available source of information with regard to water uses in each River Basin District is a master plan study conducted by the Ministry of Environment which only contained general socioeconomic information. More, financial data from the drinking and irrigation water companies were not always available since their legal status does not oblige them to report their economic elements. This was especially true for smaller companies operating in small towns. Finally the implementation of the Article 4 which defines the environmental objectives per RBD was not completed at the time of the study and thus the environmental quality assessment had to be based on approximations from existing studies. The agricultural census was not organized per RBD as well. Therefore information with respect to cultivations and water demand were approximated. The European Commission has taken Belgium, Denmark, Greece and Portugal to court over their failure to comply with EU water legislation and submit plans for managing their river basins. These plans should have been adopted by 22 December 2009 the latest.

4 Total Cost of Water Services

The most important economic concept that the Directive introduces, is that of water resources management based on the recovery of the total economic cost of water services such as freshwater provision to domestic uses and irrigation, urban wastewater collection and treatment by the Sewerage Services and recycled water supply to irrigation. According to Article 9 the total economic cost of water includes the financial cost of water companies (including costs of investments, operation and maintenance costs and administrative costs), but also the environmental and resource costs. The environmental cost reflects social welfare losses associated with water quality deterioration, caused by the water uses, while the resource cost represents additional costs that are, or will be, needed in order to cover water demand under water deficits due to the overexploitation of available water resources. In this respect Article 9 clearly states that MS 'shall take account of the principle of recovery of the costs of water services, including environmental and resource costs, having regard to the economic analysis conducted according to Annex III, and in accordance in particular with the polluter pays principle'. Table 1.2 illustrates the disaggregation of the total cost of water services. Moreover, the WFD also states that the cost recovery of water services should be analyzed for different water uses, which should be at least disaggregated into households, industry and agriculture.

---

[2] Since 2009 the name of the ministry is changed to Ministry of Environment, Energy and Climate Change.

Table 1.2 Total cost of water

Total economic cost of water services	Financial cost	Cost of providing and administering water services. Includes: capital cost, operation cost, maintenance cost, administrative cost
	Environmental cost	Environmental cost represents the costs of damage that water uses impose on the environment and ecosystems and those who use the environment (e.g., a reduction in the ecological quality of aquatic ecosystems or the salinization and degradation of productive soils)
	Resources cost	Resource cost represents the costs of foregone opportunities which other uses suffer due to the depletion of the resource beyond its natural rate of recharge or recovery (e.g., linked to the over-abstraction of groundwater)

Overall, with respect to the cost analysis of the water uses the following steps are recommended (MoEPPW 2006, pp. 233–234):

- The identification of significant pressures and impacts, which derives from the analysis of pressure and impacts
- Geographical and qualitative assessment of the various water uses (agricultural, industrial, domestic) in the regions of each RBD
- Estimation of socio-economic significance of water uses
- Identification of protected areas with species that present high economical value

5 The Water Situation in Greece and the Potential of the Timely Implementation of the EU WFD

In Greece, the WFD has been transposed into the national legislation with Law 3199/2003 (MoEPPW 2003). The country occupies a total area of 131,957 km^2 divided into 14 RBDs as presented in Fig. 1.3, out of which 5 are international sharing water courses with Albania, FYROM and Bulgaria to the north and Turkey to the east. Furthermore the country is divided into 45 River Basins (Fig. 1.4). At this point, since our case study for the chapters to follow is Asopos (RB), it should be noted that Asopos RB (GR25, Fig. 1.4) is part of the Water District (GR 07) of East Sterea Ellada (Fig. 1.3).

In Greece water supply is viewed as a public service and it is mainly the municipalities which are responsible for water supply, wastewater collection, treatment and disposal. In the largest cities of the country, Athens and Thessaloniki, owned companies (non profit making corporations), controlled by the Ministry of

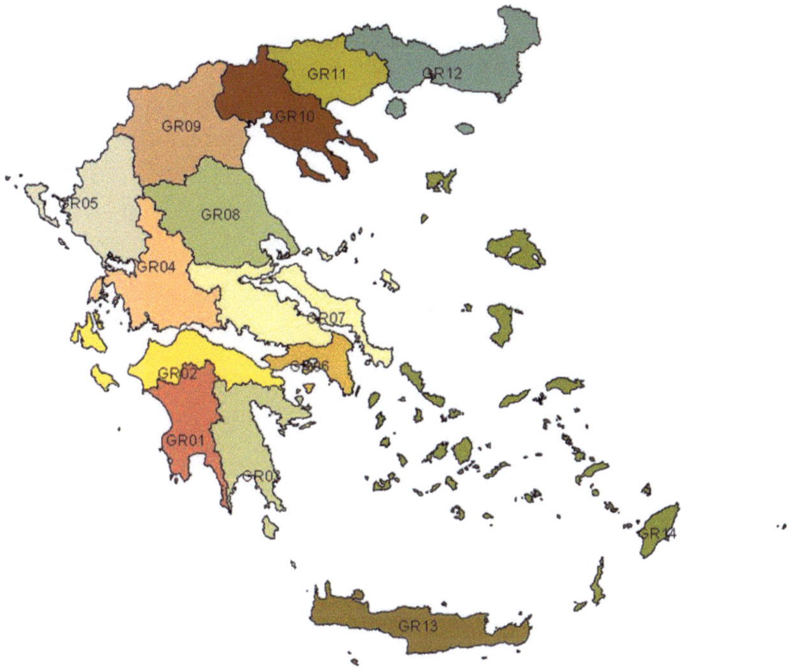

Fig. 1.3 Greek river basin districts. *01*: West Peloponnese, *02*: North Peloponnese, *03*: East Peloponnese, *04*: West Sterea Ellada, *05*: Epirus, *06*: Attica, *07*: East Sterea Ellada, *08*: Thessaly, *09*: West Macedonia, *10*: Central Macedonia, *11*: East Macedonia, *12*: Thrace, *13*: Crete, *14*: Aegean Islands (Source: http://www.minenv.gr/nera/ (WFD Article 3 report - Greek maps))

Environment Physical Planning and Public Works, own and operate the treatment plants. In the other cities of more than 10,000 inhabitants water supply is managed by municipal companies – operating as private enterprises DEYA (Municipal Enterprise for Water Supply and Sewerage) but owned by the municipalities. The pricing policy is determined by each DEYA on the basis of their cost and is approved by the Municipal Council. Based on the economic elements on DEYA the mean price per cubic meter of water in Greece is estimated at €1.27.

As far as the agricultural sector is concerned the 40 % of Greece's irrigation needs are covered by the 404 operating Local Irrigation Companies, which are responsible for the abstraction and distribution of water. The construction of major irrigation plants is undertaken by the General Irrigation Companies operating in 10 river basin districts. Prices are set by irrigation companies based on private cost criteria. The mean price per irrigated thousand square meters in Greece is €13.73.

Table 1.3 presents the overall socio-economic characterization of the RBDs. In particular, for each river basin it is reported the percentage of participation of

Fig. 1.4 Greek river basins (Source: http://www.minenv.gr/nera/ (WFD Article 3 report – Greek maps))

Table 1.3 Socio-economic identity of the RBDs

River basin district	Primary sector (% GDP)	Secondary sector (% GDP) (processing)	Tertiary sector (% GDP)
1. West Peloponnese	42.2	17.7 (9.0)	40.1
2. North Peloponnese	18.6	34.3 (22.6)	47.1
3. East Peloponnese	33.7	24.7 (3.4)	41.6
4. West Sterea Ellada	34.8	18.8 (6.1)	46.4
5. Epirus	26.3	19.5 (6.7)	54.2
6. Attica	2.1	26.5 (18.6)	71.4
7. East Sterea Ellada	24.1	43.0 (27.4)	32.9
8. Thessaly	33.5	26.1 (10.6)	40.3
9. West Macedonia	25.9	37.0 (7.6)	37.1
10. Central Macedonia	14.9	30.5 (14.4)	54.6
11. East Macedonia	27.6	32.5 (8.2)	39.8
12. Thrace	34.7	29.4 (12.7)	35.9
13. Crete	28.7	16.3 (3.8)	55.0
14. Aegean Islands	15.2	20.2 (5.6)	64.6

Table 1.4 Economic analysis of the most important water uses and pressures in each RBD

River basin district	Population (2001)	Area (km^2)	Demand for supply (hm^3 /year)	Demand for irrigation (hm^3 /year)	Demand for industry (hm^3/year)
1. West Peloponnese	331,180	7,301	23	201	3
2. North Peloponnese	615,288	7,310	36.7	395.3	3
3. East Peloponnese	288,285	8,477	22.1	324.9	0.03
4. West Sterea Ellada	312,516	10,199	22.4	366.5	0.35
5. Epirus	464,093	10,026	33.9	127.4	1
6. Attica	3,737,959	3,207	400	99	1.5
7. East Sterea Ellada	577,955	12,341	41.6	773.7	12.6
8. Thessaly	750,445	13,377	69	1,550	0.054
9. West Macedonia	596,891	13,440	43.7	609.4	30
10. Central Macedonia	1,362,190	10,389	99.8	527.6	80
11. East Macedonia	412,732	7,280	32	627	0.321
12. Thrace	404,182	11,177	27.9	825.2	11
13. Crete	601,131	8,335	42.33	320	4.1
14. Aegean Islands	508,807	9,103	37.19	80.20	1.24

each sector of economy (primary, processing, and services) to the formation of the total GDP of the river basin. Table 1.4[3] presents the most important water uses in each RBD. From these tables it is apparent that East Sterea which includes Asopos RB presents a considerable contribution of the secondary sector in GDP and a considerable demand of water use for irrigation and industry.

According to the information/data provided by the National Management Program of water inventory of the Hellenic Ministry of the Environment, Physical Planning and Public Works, the river basins are distinguished according to the conditions of water quality to:

1. Good
2. Moderate
3. Bad

This distinction has been done following the measurements in each river basin regarding the concentrations of NO_3, P and NH_4. The concentration for each pollutant is characterized as Low, Moderate or High according to the levels of the pollutant factor.[4] The water quality in a river basin is characterized as Good if the majority of the measurements for all pollutants indicate Low Concentration, Moderate if the

[3] Report on the implementation of Article 5 of the WFD (2008).

[4] Low Concentration: **P**<0.17mg/l, **NO_3-N**<5 mg/l, **NH_4-N**<0.04 mg/l
Moderate Concentration : 0.17 mg/l <**P**<0.31 mg/l, 5 mg/l <**NO_3-N**<11 mg/l, 0.04 mg/l <**NH_4-N**<1
High Concentration: **P**>0.31 mg/l, **NO_3-N**>11, **NH_4-N**>1

Table 1.5 Overall condition of river quality

River basin district	Concentration NO$_3$ Low	Moderate	High	P Low	Moderate	High	NH$_4$ Low	Moderate	High	Total condition
1. West Peloponnese	2	0	1	2	0	0	0	0	0	Good
2. North Peloponnese	1	0	0	1	0	0	0	1	0	Good
3. East Peloponnese	1	1	0	2	0	0	1	0	0	Good
4. West Sterea Ellada	11	0	0	10	2	1	0	9	0	Good
5. Epirus	7	0	0	8	0	1	3	4	0	Good
6. Attica	1	0	0	2	0	0	1	0	0	Good
7. East Sterea Ellada	3	0	2	1	2	0	0	3	0	Moderate
8. Thessaly	3	1	0	2	2	0	0	4	0	Moderate
9. West Macedonia	12	1	1	10	3	10	0	11	5	Bad
10. Central Macedonia	7	0	0	2	1	7	0	7	0	Moderate
11. East Macedonia	7	0	0	3	3	3	0	7	0	Moderate
12. Thrace	8	9	1	0	5	12	0	13	4	Bad
13. Crete	4	0	0	4	0	0	4	0	0	Good
14. Aegean Islands	–	–	–	–	–	–	–	–	–	Good

majority of measurements indicate Moderate Concentration and finally water quality is characterized as Bad if the majority of the measurements for all the considered pollutants indicate High Concentration. Table 1.5[5] presents the total number of the available measurements per pollutant and concentrations as well as the final condition of the river basin. Regarding the East Sterea Ellada RBD it demonstrates an overall "moderate" condition of river quality considering the specific examined parameters. It should be also reminded that Asopos RB is only part of East Sterea Ellada RBD and its specific condition will be presented in the following two chapters which focus more on this particular basin characterization.

Thc following section analyzes the calculation of the financial, environmental and resources cost that determine the total cost of water services that should be taken into account in the design of future pricing policies to ensure sustainable water resources management in line with the provisions of the Directive.

[5] Report on the implementation of Article 5 of the WFD (2008).

6 Methodologies for the Calculation of Total Economic Cost in Greece

6.1 Financial Cost

The financial cost of water services includes operational, administrative, maintenance costs of existing infrastructure and investment cost for the enterprises of drinking water supply and sewerage and the irrigation water companies. The relevant data, for the calculations under this study, were collected from the enterprises' annual published financial reports of the most recent 5 years. For the RBDs where financial data were not available for all enterprises, the total financial cost was approximated assuming for the remaining enterprises the Greek mean financial cost per enterprise and aggregating overall operating enterprises. Financial costs for domestic and agricultural water supply in each RBD are presented in Table 1.6.

6.2 Environmental Cost

The environmental cost refers to the cost associated with water quality depletion and thus the subsequent limitation of water resources' capacity to provide goods and services which can be translated to value for people. Values from water resources include both values associated with the direct use of water for drinking, irrigation for agriculture and recreation, but also non-use values relating to nutrient retention, flood control and protection, biodiversity and bequest and aesthetic purposes among

Table 1.6 Financial cost per RBD

	Financial cost (€)	
River basin district	Domestic (€/hm^3)	Irrigation (€/ha)
1. West Peloponnesos	4,108,662	27.3
2. North Peloponnesos	4,612,819	14.6
3. East Peloponnesos	6,895,954	253
4. West Sterea Ellada	4,762,739	334.4
5. Epirus	5,684,518	319.2
6. Attica	833,711	13
7. East Sterea Ellada	3,378,763	10.07
8. Thessaly	6,850,916	63.9
9. West Macedonia	3,934,249	33.5
10. Central Macedonia	2,091,853	53
11. East Macedonia	5,193,781	95.7
12. Thrace	2,746,149	28.6
13. Crete	5,258,926	33.8
14. Aegean Islands	9,530,520	10.3

others (Birol et al. 2006) To calculate the environmental damage arising from water supply or discharge, a variety of valuation techniques developed by economists can be applied which are generally classified as revealed preference techniques (see for example Braden and Kolstad 1991) and stated preference techniques (see for example Adamowicz et al. 1998). The first take into account observable market information which can be adjusted and used for revealing the individual's preferences while in Stated Preference approaches the market for the good is 'constructed' through the use of questionnaires and respondents are asked to state their willingness to pay for an improvement.

In the present study, in order to calculate the environmental cost a Benefits Transfer approach (Kirchhoff et al. 1997; Desvousges et al. 1992) is implemented. This approach allows values from existing studies to be transferred to policy sites of interest after correcting for certain parameters. The mean value from an extensive literature of valuation studies which apply stated preference methods to elicit individual valuations for water quality amelioration was thus adopted under this study after proper adjustment to reflect Greek-specific socioeconomic characteristics.[6] A single value transfer was judged as inappropriate under this study due to the great heterogeneity between the different RBD with regards to both environmental and socioeconomics conditions. We argue that a mean value from studies conducted in different areas with different cultural, environmental and socioeconomic backgrounds can isolate case-study specific characteristics and thus can better approximate the value to be adopted for this study. Monetary estimates from the considered studies capture both values associated with the direct use of water but also non-use values relating to the existence of water ecosystems.

RBDs were initially classified according to the pollution loads identified. The evaluation of surface and groundwater quality characteristics for the classification was based on monitoring data under existing studies (MoEPPW 2007). Accordingly, the environmental cost was approximated with the welfare loss due to resources depletion and degradation as elicited in existing valuation studies after proper adjustment. Welfare estimates in the considered studies are reported as WTP per individual and they were then aggregated over the population of each RBD. The environmental cost was then disaggregated into different uses based on the pollution loads that are attributed to each use. The contribution percentage of the use in the total pollution loads identified in the RBD $\left(x = \frac{Polution\ load\ attributed\ to\ a\ use}{total\ pollution\ load\ in\ the\ RBD} \right)$ was then used to approximate the environmental cost for this use (environmental cost of a use= x*total environmental cost in the RBD). An ideal approach to calculate the environmental cost would involve original valuation studies to be carried out in

[6]Ahmad et al. (2005), Basili et al. (2006), Bateman et al. (2004), Birol and Koundouri (2008b), Birol et al. (2006), Brouwer and Bateman (2005), Crandall (1991), Crutchfield et al. (1997), Day and Mourato (2002), Farber and Griner (2000), Forster (1985), Georgiou et al. (2000), Green et al. (1993), Green and Tunstall (1991), Hanley (1991), Jordan and Elnagheeb (1993), Lindhjem (1998), Oglethorpe and Miliadou (2000), Mitchell and Carson (1984), Ozdemiroglu et al. (2004), Poe and Bishop (1992), Whitehead and Groothuis (1992) and Koundouri (2007, 2010).

Table 1.7 Environmental cost per RBD

River basin district	Environmental cost (€)
1. West Peloponnesos	0
2. North Peloponnesos	0
3. East Peloponnesos	0
4. West Sterea Ellada	0
5. Epirus	0
6. Attica	0
7. East Sterea Ellada	7,037,232
8. Thessaly	9,137,486
9. West Macedonia	14,535,598
10. Central Macedonia	16,586,149
11. East Macedonia	5,025,462
12. Thrace	9,842,713
13. Crete	0
14. Aegean Islands	0

each water body of each river basin district addressing the particular environmental problem in the area to allow for accurate welfare loss estimations. These valuation studies could either follow the states preference paradigm or implement revealed methods to derive the social value of water quality. However such an approach would require large financial resources and a more extended time horizon. Nevertheless, the benefit transfer approach presented above is widely accepted for providing, if properly adjusted, reliable costs and benefits estimates. The results of the environmental cost calculations are reported in the Table 1.7 above.

As expected the largest environmental cost is in northern Greece and specifically in Macedonia. Water quality in these regions is under severe stress since they concentrate the bulk of the remaining industrial activity in the country. On the other hand, island regions and the Peloponnese that have not developed industrial sectors have smaller environmental costs.

6.3 Resource Cost

The resource cost is a cost associated with current or future scarcity arising due to overexploitation of water resources beyond their rate of replenishment implying that resource cost is present when water demand for all uses is not covered adequately and is zero otherwise. The resource cost was calculated for the water districts of Aegean Islands, East Sterea Ellada, Thessaly and East Peloponnesos, where water demand surpasses supply as indicated by their water balance. In the literature (Koundouri 2004) resource cost is approximated by the cost of backstop technology to cover excess demand. Desalination plants are set up in many Aegean islands and thus the price of this backstop technology was used for the resource cost approximation in the relevant water district. The exploitation of other non-conventional water sources such as recycled water was the backstop technology relevant for the

Table 1.8 Resource cost per RBD

River basin district	Resource cost (€)
1. West Peloponnesos	0
2. North Peloponnesos	0
3. East Peloponnesos	3,510,184
4. West Sterea Ellada	0
5. Epirus	0
6. Attica	0
7. East Sterea Ellada	20,515,680
8. Thessaly	89,356,467
9. West Macedonia	0
10. Central Macedonia	0
11. East Macedonia	0
12. Thrace	0
13. Crete	0
14. Aegean Islands	26,792,100

water districts of East Peloponnesos and East Sterea Ellada, whereas the diversion of the river Acheloos is meant cover excessive water demand in the water district of Thessaly. Resource cost was thus the product of the excess demand times the backstop technology cost per cubic meter of water which is €1.5 m^3 for the desalination (WDD 2005), €0.5/m^3 for the recycled water (WDD 2005) and €0.818 m^3 for the Acheloos diversion (personal communication MoEPPW). Water shortages are attributed to the agricultural sector because (a) agriculture uses 80% of the available water resources (b) agriculture has been identified as the most inefficient water using sector and (c) there exist a policy priority to cover water needs of the residential sector, which constitutes 17 % of total water consumption. Accordingly, no resource cost was attributed to domestic water use. Table 1.8 reports the resource cost in each RBD.

The largest resource cost is in Thessaly. This is the primary agricultural region of Greece that faces threats on water availability. Further resource cost is also high in the Aegean Islands which experience severe water shortages due to limited rainfall especially during the summer months.

7 Assessment of Cost-Recovery Level

7.1 Recovery from Charges to Users

Once the total cost of water services is determined and the revenues of water companies are calculated an assessment of the cost- recovery level is possible. Hence we now turn to the calculation of the current level of cost recovery, based on the present pricing structure in each of the considered economic sectors. As far as the domestic sector is concerned, the water companies cost recovery derives from both potable water pricing and sewerage connection and wastewater treatment pricing. The later, is calculated as a surcharge of 80 % to the value of water consumption, with small fluctuations from region to region. Revenues from water consumption were calculated by multiplying the consumed cubic meters of water with the mean water price in each RBD whereas the sewerage expenses were inferred given the number of households in each RBD and the relevant fees set by the water companies. With respect to irrigation companies, cost recovery was calculated by multiplying the irrigation needs with the mean irrigation water price per RBD.

7.2 Estimating the Cost-Recovery Level

When both the total economic cost and the revenues from charges to users were identified for each use in every RBD the cost recovery level was calculated as:

$$Cost\ Recovery\ Level = \frac{Recovery}{Total\ Economic\ Cost}$$

The mean cost recovery level per RBD in Greece was found at 59.18 %. In general, the revenues of water and sewerage services providers are not even sufficient for financial cost recovery. It is worth noting that the recovery level in agriculture is even lower compared to domestic water use, stressing the need for policy measures to address sustainable and efficient water resources management in this sector. The problem is particularly prominent in the district of Aegean Islands where irrigation cost recovery level is 1.78 %. The results of the economic analysis regarding the cost recovery level are summarized in Table 1.9.

The above estimates should be regarded as broad estimates of the true recovery level, the calculation of which would be expensive and extremely difficult to conduct in the short term. The second best approach pursued based on benefit transfers and reasonable assumptions, however allows for valuable conclusions to be reached regarding the limitations of the pricing policies which fail to reflect the true value of the resource and efficiently allocate it to competing demands. Given that most MS according to the commission (COM (2007) 128 final) provided incomplete reports for the economic analysis regarding the recovery level, the methodology followed in this study can assist future attempts to fully comply with the EU reporting requirements. Besides, only five MS having supplied information on cost recovery have taken into consideration in the analysis environmental and resource costs (see Fig. 1.3). In this respect, the Greek case study can provide useful guidance.

Because of the lack of information, it is also not possible to give an average on the percentage of cost recovery across the EU. When available, cost-recovery levels vary significantly (WWF 2006). MS that have provided information on households have indicated a cost recovery rate of services for households between 70 % and 100 %. For industry, the MS providing information reported a cost recovery rate between 40 % and 100 %. For agriculture the cost recovery rate is reported to vary between 1 % and 100 % (Fig. 1.5).

Table 1.9 Cost recovery level in each RBD

	Cost recovery level (%)		
River basin district	Domestic	Irrigation	Total
1. West Peloponnesos	62.21	11.44	50.54
2. North Peloponnesos	77.31	19.41	68.22
3. East Peloponnesos	37.89	15.66	34.18
4. West Sterea Ellada	61.29	14.28	46.19
5. Epirus	71	22.44	68.11
6. Attica	108.14	21.30	106.13
7. East Sterea Ellada	75.1	15.98	57.61
8. Thessaly	33.66	6.38	29.82
9. West Macedonia	53.55	41.05	51.71
10. Central Macedonia	86.58	12.04	78.27
11. East Macedonia	79.39	27.38	70.74
12. Thrace	103.29	11.05	78.28
13. Crete	49.67	56.25	50.91
14. Aegean Islands	42.94	1.78	37.84

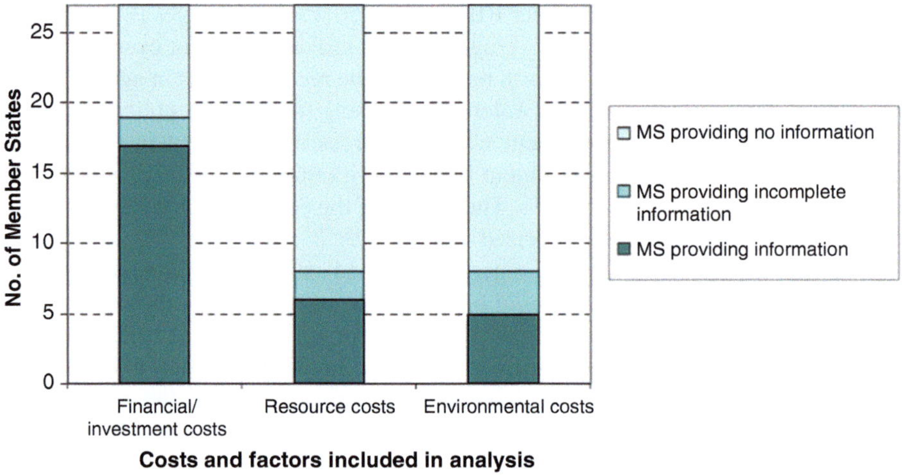

Fig. 1.5 MS providing information on costs of water services (Source: (COM (2007) 128 final))

8 Discussion and Conclusions

The WFD was formulated to address the weaknesses of previous water-related directives by adopting an integrated water management approach. Given the increasing pressures on the quality and quantity of water resources, the EU has established an effective legislative instrument in the form of the WFD. The innovation of the WFD is that, for the first time a directive proposes economic principles and measures as basic instruments for the achievement of specific environmental objectives. According to Kallis and Butler (2001) the main strengths of the Directive apart from the broader and integrated ecosystem approach is that it introduces changes with respect to institutions, planning and information processes, and the 'user-pays' approach, but importantly sets a concrete standard of no further deterioration for any water. However, the economic analyses of most MS are incomplete and this is one of the major shortcomings in the WFD implementation so far. This concerns in particular the calculation of water services cost recovery accounting for environmental and resource costs (COM (2007) 128 final).

At the same time, serious concerns about the success of the Directive have been expressed. For example, the WFD requires that charges for water services should adopt the principle of full cost recovery in accordance with the polluter pays principle in order to provide incentives for water use efficiency. However, it is expected to be quite challenging in a number of MS where water in the domestic and agricultural sectors is subsidized (Spain, Greece, and Portugal) or water pricing is completely absent (Ireland). In the latter case, the political cost of asking households to pay for environmental improvements when sources of diffuse pollution are not fully checked is expected to be high. Furthermore, pricing mechanisms imply 'benefit pricing' based on willingness to pay and there is a fear of discriminatory practices from the side of profit seeking suppliers (Morris 2004).

Regarding assessment of "disproportionality", it has been argued that whether or not costs are considered disproportionate is highly arbitrary and subjective (European Communities 2002; Brouwer 2008) as it remains to be answered (i) what is an acceptable cost level in relation to the expected environmental benefits for example, being a maximum of two, three or four times the expected (monetary) benefits; and (ii) what is the acceptability of this decision to those who bear the financial burden (Brouwer 2008). It has been also noted that it is highly questionable whether policy makers and society as a whole are willing to pay the relevant investment sums without any further justification as to their socio-economic benefits (Brouwer 2008), while Brouwer and Pearce (2005) argue that European legislation such as the WFD introduces 'asymmetric property rights' assigning higher weights to environmental benefits compared to the social costs involved.

Kallis and Butler (2001) express a fear that ambiguity of terms especially related to derogations coupled with the high costs involved and the lack of a clear-cut legal mandate to achieve the status objectives may undermine the effectiveness of the policy as unwilling MS may exploit legislative loopholes to avoid implementation. Finally, Carter and Howe (2006) argue that the WFD is an ambitious piece of legislation and its key objective to achieve good water status in most of Europe's waters is not expected to be achieved in the short term (by 2015).

In this chapter we present a rapid-appraisal approach we followed in Greece for the estimation of the cost-recovery level in its 14 RBDs including the RBD of Asopos (East Sterea Ellada). The approach seeks to provide guidance and assist policymakers and researchers in other RBD to proceed with the assessment of the cost-recovery level. Results reveal that the recovery level in Greece is extremely low in all water sectors. In the majority of cases not even the financial cost of water companies in both domestic and irrigation water sectors is covered. Environmental and resource costs associated with water uses are not taken into consideration in the determination of the pricing policies resulting in very low pricing of water for households and industry and almost free of charge water provision in agriculture, particularly in the presence of many illegal boreholes, as is apparent from the very low recovery level.

Authors, however, acknowledge the limitations of this study. A comprehensive and complete analysis of all pressures and impacts and a detailed assessment of the cost recovery in each river basin as required by the Directive were not possible in this study due to the severe lack of data and very limited time frame. Scale and time inconsistencies of the available data have also been identified and approximations were thus used. Detailed and site specific data were rare and consequently results and policy recommendations should be considered conditional on these limitations. However, results under this 'quick appraisal' clearly highlight the need for reforms in the current pricing policy to adequately address full cost recovery of water services according to the polluter pays principle as required by the WFD. To address this challenge, reforms in the institutional framework covering water resources management are also clearly needed. Overlapping responsibilities between competent authorities pose a serious constraint in efficient and sustainable water resources management. All these considerations should be taken into account in the preparation of the package of measures to reach good water status and ensure full recovery of the cost of water services as required under Article 11 of the WFD.

References

Adamowicz, W., Boxall, P., Williams, M., & Louviere, J. (1998). Stated preference approaches for measuring passive use values: Choice experiments and contingent valuation. *American Journal of Agricultural Economics, 80*, 64–75.

Ahmad, J., Goldar, B., & Misra, S. (2005). Value of arsenic-free drinking water to rural households in Bangladesh. *Journal of Environmental Management, 74*, 173–185.

Basili, M., Di Matteo, M., & Ferrini, S. (2006). Analysing demand for environmental quality: A willingness to pay/accept study in the Province of Siena (Italy). *Waste Management, 26*(3), 209–219.

Bateman, I. J., Cole, M., Cooper, P., Georgiou, S., Hadley, D., & Poe, G. L. (2004). On Visible choice sets and scope sensitivity. *Journal of Environmental Economics and Management, 47*, 71–93.

Birol, E., & Koundouri, P. (2008a). *Choice experiments in Europe: Economic theory and applications*. Edward-Elgar Publishing, Wally Oates and Henk Folmer's 'New Horizons in Environmental Economics' Series. ISBN: 9781845427252.

Birol, E., & Koundouri, P. (2008b). *Choice experiments informing environmental policy: A European perspective*. Edward-Elgar Publishing, Wally Oates and Henk Folmer's 'New Horizons in Environmental Economics' Series. ISBN: 9781845427252 (337pages).

Birol, E., Karousakis, K., & Koundouri, P. (2006). Using economic valuation techniques to inform water resources management: A survey and critical appraisal of available techniques and an application. *Science of the Total Environment, 365*, 105–122.

Braden, J. B., & Kolstad, C. D. (Eds.). (1991). *Measuring the demand for environmental quality*. Amsterdam: North-Holland.

Brouwer, R. (2008). The role of stated preference methods in the Water Framework Directive to assess disproportionate costs. *Journal of Environmental Planning and Management, 51*(5), 597–614.

Brouwer, R., & Bateman, I. J. (2005). Temporal stability and transferability of models of willingness to pay for flood control and wetland conservation. *Water Resources Research, 41*(3), W03017.

Brouwer, R., & Pearce, D. W. (2005). Introduction. In R. Brouwer & D. W. Pearce (Eds.), *Cost benefit analysis and water resources management*. Cheltenham: Edward Elgar.

Brouwer, R., Barton, D. N., Bateman, I. J., Brander, L., Georgiou, S., Martín-Ortega, J., Pulido-Velazquez, M., Schaafsma, M., & Wagtendonk, A. (2009). *Economic valuation of environmental and resource costs and benefits in the Water Framework Directive: Technical guidelines for practitioners*. Amsterdam: Institute for Environmental Studies, VU University.

Carter, J., & Howe, J. (2006). The Water Framework Directive and the Strategic Environmental Assessment Directive: Exploring the linkages. *Environmental Impact Assessment Review, 26*, 287–300.

Commission of the European Communities. (2007). *Towards Sustainable Water Management in the European Union. First stage in the implementation of the Water Framework Directive 2000/60/EC*. Accompanying document to the Communication from the Commission to the European Parliament and the Council. COM (2007) 128 final.

Commission of the European Communities (CEC). (2000). *Directive of the European Parliament and of the Council establishing a framework for Community action in the field of water policy*. 1997/0067 (COD), C5-0347/00.

Common Implementation Strategy for the Water Framework Directive (CIS). (2000/60/EC). *Moving to the next stage in the Common Implementation Strategy for the Water Framework Directive Progress and work programme for 2005 and 2006* AS AGREED BY THE WATER DIRECTORS 2/3 December 2004.

Crandall, K. B. (1991). *Measuring the economic benefits of riparian areas*. Masters thesis, University of Arizona.

Crutchfield, S. R., Cooper, J. C., & Hellerstein, D. (1997). *Benefits of safer drinking water: The value of nitrate reduction*. Washington, DC: Food and Consumer Economics Division, Economics Research Service, U.S. Department of Agriculture.

Day, B., & Mourato, S. (2002). Valuing river water quality in China. In D. Pearce, C. Pearce, & C. Palmer (Eds.), *Valuing the environment in developing countries: Case studies*. Edward Elgar: Cheltenham/Northampton.

Desvousges, W. H., Naughton, M. C., & Parsons, G. R. (1992). Benefit transfer: conceptual problems in estimating water quality benefits using existing studies. *Water Resources Research, 28*(3), 675–683.

European Communities. (2002). Guidance document No. 1, *Economics and the environment – the implementation challenge of the Water Framework Directive. A guidance document*. Common Implementation Strategy for the Water Framework Directive (2000/60/EC). Brussels: European Commission. Available at http://forum.europa.eu.int/Public/irc/env/wfd/library

Farber, S., & Griner, B. (2000). Valuing Watershed Quality Improvements Using Conjoint Analysis. *Ecological Economics, 34*, 63–76.

Forster, B. A. (1985). Economic impact of acid deposition in the Canadian Aquatic Sector. In D. F. Adams & W. W. Page (Eds.), *Acid Deposition*. New York: Plenum Publishers.

Georgiou, S., Bateman I., Cole M., & Hanley, N. (2000). *Contingent ranking and valuation of river water quality improvements: Testing for scope, sensitivity, ordering and distance decay effects*. CSERGE Working Paper GEC 2000–18, UK Economic and Social Research Centre.

Green, C. H., & Tunstall, S. M. (1991). The evaluation of river quality improvements by the contingent valuation method. *Applied Economics, 23*, 1135–1146.

Green, C., Tunstall, S., Herring M., & Sawyer, J. (1993). *Customer preference and willingness to pay for selected water and sewerage services*. Report to the Office for Water Services.

Hanley, N. (1991). The economics of nitrate pollution in the UK. In N. D. Hanley (Ed.), *Farming and the countryside: An economic analysis of external costs and benefits*. Oxford: CAB.

Hanley, N., & Black, A. R. (2006). Cost-benefit analysis and the Water Framework Directive in Scotland. *Integrated Environmental Assessment and Management, 2*(2), 156–165.

Hellenic Republic. Ministry of Environment Physical Planning and Public Works (MoEPPW). (2003). *The law for the management of water resources of Greece*, Law N.3199/03. Athens, Greece.

Hellenic Republic. Ministry of Environment Physical Planning and Public Works (MoEPPW). (2006). *Report on the pressures and qualitative characteristics of water bodies in the water districts of Greece and a methodological approach for further analysis*. Athens, Greece.

Hellenic Republic. Ministry of Environment Physical Planning and Public Works (MoEPPW). (2007). *National programme for the management and protection of water resources, in Greek*. Athens, Greece.

Jordan, J. L., & Elnagheeb, A. H. (1993). Willingness to pay for improvements in drinking water quality. *Water Resources Research, 29*(2), 237–245.

Kallis, G., & Butler, D. (2001). The EU water framework directive: measures and implications. *Water Policy, 3*, 125–142.

Kirchhoff, S., Colby, B. G., & LaFrance, J. T. (1997). Evaluating the performance of benefit transfer: An empirical inquiry. *Journal of Environmental Economics and Management, 33*, 75–93.

Koundouri, P. (2004). Current issues in the economics of groundwater resource management. *Journal of Economic Surveys, 18*(5), 703–740.

Koundouri, P. (2007). *Coping with water deficiency: From research to policy making* (Environment and policy series, Vol. 48). Springer: Berlin. ISBN 978-1-4020-6614–6.

Koundouri, P. (2010). *Water resources allocation: Policy and socioeconomic issues in Cyprus* (Environment and policy series). Dordrecht: Springer.

Koundouri, P., & Remoundou, K. (2009). Introduction to the book. In P. Koundouri (Ed.), *The use of economic valuation in environmental policy: Providing research support for the implementation of the EU water policy under AquaStress* (Routledge explorations in environmental economics). London/New York: Routledge/Taylor and Francis Group.

Lindhjem, H. (1998). *Willingness to pay for improved water quality of Steinsfjorden – Betalingsvillighet for en bedret vannkvalitet i Steinsfjorden*. Oslo: The University of Oslo.

Mitchell, R. C., & Carson, R. T. (1984). *Willingness to pay for national freshwater quality improvements*. Washington, DC: Resources for the Future.

Morris, J. (2004). *Economics of the water framework directive: purpose, principles and practice*. Environmental Economics Conference, 26 March, The Royal Society, London.

Oglethorpe, D. R., & Miliadou, D. (2000). Economic valuation of the non-use attributes of a wetland: A case-study for Lake Kerkini. *Journal of Environmental Planning and Management, 43*(6), 755–767.

Ozdemiroglu, E., Newcombe, J., Mourato, S., Atkinson, G., & deGaris, Y. (2004). *The value of a Tidier Thames: WTP to reduce sewage overflows, applied environmental economics conference*, The Royal Society, London.

Poe, G. L., & Bishop R. C. (1992). *Measuring the benefits of groundwater protection from agricultural contamination: Results from a two-stage contingent valuation study* (Agricultural Economics Staff Paper No. 341). Madison: University of Wisconsin.

Ramsar Convention. (1996). *Strategic plan 1997–2002*. Gland: Ramsar Convention Bureau.

Report on the implementation of Article 5 of the WFD, Hellenic Ministry of the Environment, Physical Planning and Public Works, Athens March 2008, Prepared by RESEES – REsearch on Socio-Economic and Environmental Sustainability – team available at: http://www.aueb.gr/users/koundouri/resees/en/aswposprojen.html (In Greek).

UNEP (United Nations Environment Program). (2005). *Vital water statistics*. (http://www.unep.org/vitalwater/).

Water Development Department, Republic of Cyprus. (2005). *Eu summary report articles 5&6*. Available at: http://www.cyprus.gov.cy/moa/wdd/Wdd.nsf/All/B8D7262CBFCC9AF8C225711E00303F5A/$file/Page1-20.pdf. Accessed Dec 2008.

Whitehead, J. C., & Groothuis, P. A. (1992). Economic benefits of improved water quality: A case of North Carolina's Tar-Pamlico River. *Rivers, 3*(3), 170–178.

WWF. (2006) *EU Water Policy: Making economics work for the environment: Survey of the economic elements of the Article 5 report of the EU Water Framework Directive*, available at http://assets.panda.org/downloads/eu_water_policy___may_2006.pdf

Chapter 2
Introducing the Case Study, the Asopos River Basin in Greece

P. Koundouri, N. Papandreou, M. Stithou, A. Mousoulides, Y. Anastasiou, M. Mousoulidou, A. Antypas, T. Mavrogiorgis, and K. Vasiliou

Abstract The case study area is presented in this chapter. The study area comprises the river and estuary of Asopos and the lagoon of Oropos northeast of Attica. Along the Asopos river human activities, mainly agriculture and industrial take place. Industrial and agricultural effluents pollute the river, the aquifers and the soils of the area, making it a negative example of the impacts to humans and environment that arise from non sustainable use of natural resources. The chapter starts by presenting the geographical distribution of the river system and the water bodies of the catchment. The chapter presents also a review of the related legislation that has contributed to its current situation. Then the current water status of the catchment is described, while the pressures and related impacts in the catchment are examined. Afterwards the focus is on the social and environmental functions and values of Asopos River and Oropos Lagoon. Identifying the main social, economic and environmental impacts of degradation in the area provides the background for the analysis that follows in the subsequent chapters.

Keywords Asopos River Basin • Oropos Lagoon • Institutional framework • Water quality • Wastewaters

---

P. Koundouri (✉) • M. Stithou • Y. Anastasiou • M. Mousoulidou
A. Antypas • T. Mavrogiorgis • K. Vasiliou
Athens University of Economics and Business, Athens, Greece
e-mail: pkoundouri@aueb.gr; stithoumavra@yahoo.com; yanastasiou@aueb.gr; mmousoulidou@aueb.gr; aantypas@unipi.gr; tmavrogiorgis@aueb.gr; kvasiliou@aueb.gr

N. Papandreou
Andreas G. Papandreou Foundation, Athens, Greece
e-mail: tulayev@gmail.com

A. Mousoulides
Athens University of Economics and Business, Athens, Greece

Intercollege Larnaca, Larnaca, Cyprus
e-mail: mousoulides.a@adm.lar.intercol.edu

P. Koundouri and N.A. Papandreou (eds.), *Water Resources Management Sustaining Socio-Economic Welfare*, Global Issues in Water Policy 7,
DOI 10.1007/978-94-007-7636-4_2, © Springer Science+Business Media Dordrecht 2014

1 Presentation of the Asopos River Basin

Asopos RB is part of the Water District (07) of East Sterea Ellada (Fig. 2.1). As it has been reported in MoEPPW (2006, p. 81): "Water District (07) of East Sterea Ellada occupies a total area of 12,341 km^2 and consists of the Prefecture of Evoia, major parts of the Prefectures of Fthiotida (83.1 %), Voiotia (98.5 %) and Fokida (41.9 %) and smaller parts of the Prefectures of Magnisia (14.9 %) and Attica (7.2 %). The main river basins of Water District (07) are: River basin of Voiotikos Kifissos River, River basin of Sperchios River and River basin of Asopos River. Furthermore lakes Iliki and Paralimni are significant water bodies located in the Water District (07)."

Asopos RB covers a total surface area of 450 km^2 (East Attica and Viotia Districts), and extends to Evoikos Gulf. Its sources are in Elikona but along its flow other streams from the mountains of Parnitha and Dervenochoria also contribute. Asopos has a total length of 57 km and flows through Asopia, the plain of Thiva, Oinofita, Schimatari, Sikamino and Oropos and it finally reaches North Evoikos Gulf. River's annual runoff is 70 hm^3. Figure 2.2 shows the distribution of Asopos RB highlighted in blue. A population of approximately 70,575 citizens is residing in the broad river area, which constitutes the largest industrial area in Greece, situated only 60 km north of Athens.

Fig. 2.1 Water District (07) – East Sterea Ellada (Source: MoEPPW 2006)

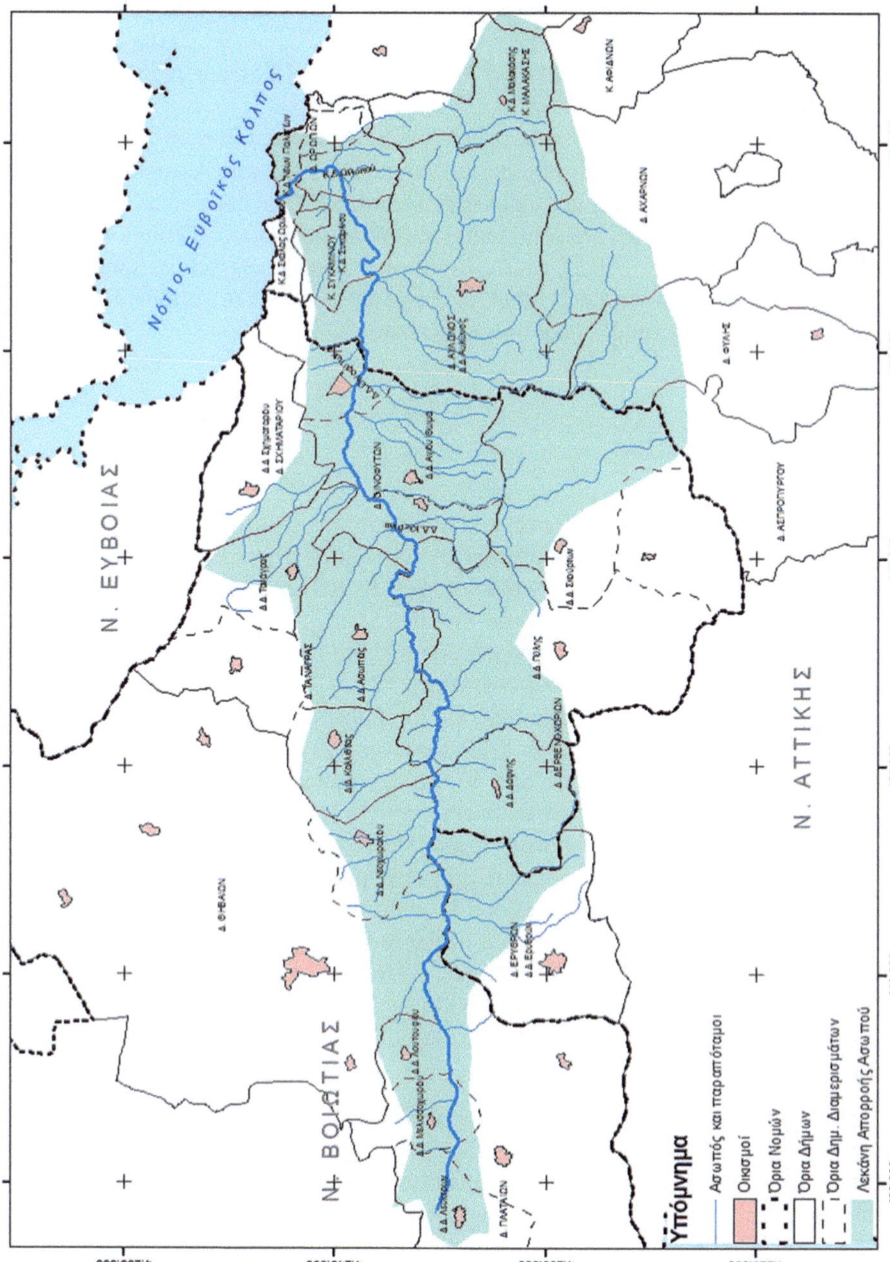

Fig. 2.2 Asopos RB – Administrative borders (Source: Apostolopoulos, 2010)

Administratively in that particular RB there are eleven Towns/Municipalities which belong in two prefectures, that of Viotia and Attiki. These eleven Towns/ Municipalities are the following: Plataion, Thivon, Tanagras, Dervenochorion, Oinofiton, Schimatariou, Avlona, Oropion, Erithron, Malakasa and Sikaminou. The last five Towns/Municipalities are a part of Attiki Prefecture while the rest form part of Viotia Prefecture. The included Towns/Municipalities are highlighted in pink in Fig. 2.2.

Finally, regarding the geotectonic outline of the Assopos valley in Economou-Eliopoulos et al. (2011, p. 40) it is reported that: "The Neogene Assopos basin is mainly composed of Tertiary and Quaternary sediments. The uppermost horizons, dominated in the studied area, are continental sediments consisting of conglomerates with small intercalations of marl, marly limestone, metaclastic schist, sandstone, clays and flysch. At the lowest parts there are alternations of lake yellow marls and marl limestones hosting small black lignite horizons."

2 Related Legislation

Before, presenting the water quality related information and the identified pressures in the study area it is important to present the institutional framework that had its impact on the current situation. In 1969 with Presidential Decree it was allowed the establishment of industrial units in the broader area of Oinofita. However, this decision did not foresee their way of operation and did not specify limits related to the industrial activities. Hence, what characterized the location of those industries in the region was the lack of any planning and monitoring system for the control of the area's industries. Ten years later in 1979, the river was appointed by the Regional Authorities (Prefectures of Attica and Viotia) as receiver of industrial sewage however, without establishing any monitoring system for the enforcement of terms and conditions under which the industries would discharge their waste in the river. Although today these terms are obsolete and in opposition with more recent laws and regulations, the treaty is still in effect. In addition, more licenses were granted for the operation of industries in the area of the region where the construction was forbidden under the 1969 Presidential Decree. As a result, the river and groundwater have been subject to long-term industrial pollution.

Regarding the implications of the related legislation the following are emphasized:

1. According to the Public Health Act/Regulation E1b/221/65 the disposal of the treated wastewater and industrial wastes is allowed either in surface water bodies after asking for Permission by the Prefect or underground (after asking Permission from the Planning Service of Northern Attiki). This Regulation is rather general and as long as the underground disposal of the treated industrial wastes are concerned, it is not referred particularly, to dangerous and toxic substances and it does not set limits for the disposal of dangerous substances.

2. The legislation for the management/disposal of the toxic-dangerous industrial wastes according to the law L. 1650/86 includes a plethora of provisions which most of them concern harmonization of the national Law with successive EU Directives. However, these provisions do not face with a single and concise way the whole issue since there are gaps, discontinuities and overlapping. Hence, although there is adequate environmental legislation relevant to the management/disposal of dangerous industrial wastes, these issues are still regulated by the Public Health Act/Regulation E1b/221/65.

In Greece, the WFD has been transposed into the national legislation with the Law 3199/2003 (MoEPPW 2003) however, till recently there was neither a Management Plan for the protection of the Asopos RB (see MoEPPW 2008) nor a Monitoring Program for its waters which should have become operational the latest until 23-12-2006 according the P.D. 51/2007 in compliance with the Directive 2000/60/EC (CEC 2000).

The substantial response of the State regarding Asopos was initiated in the beginning of 2010 where it was presented by the Minister of Environment, Energy and Climate Change a Program for the management of the environmental crisis in Asopos which aimed at the following objectives[1]:

- Ensure safe water
- Institutional safeguard of the broader area
- Effective monitoring of the pollution sources (industrial, agricultural, urban) aiming the limitation of the discharged loads
- Intensification of the controls
- Exploration of restoration measures of the environmental degradation
- The spatial planning/organization of the area

As far as the institutional and legislative framework is concerned[2]:

- The Joint Ministerial Decision was issued for the "Establishment of Environmental Quality Standards in Asopos River and Emission Limit Values for Liquid Industrial Wastes in Asopos RB", according to which (i) outdated provisions according to which the industrial units in the area could dispose their wastes in Asopos were abolished (ii) the Joint Departmental Decision according to which Asopos was defined as a disposal pipeline of industrial wastes in Evoikos has been abolished, (iii) the cost of sampling and of the laboratory analysis for the verification of complying with the relevant legislation was attributed to the controlled activity.
- The Joint Ministerial Decision was issued according to which the Asopos RB is characterized as area vulnerable to nitrate pollution and an appropriate agri-environmental program is formed in compliance with the European Directive regarding nitrates from agricultural sources.

[1] http://www.ypeka.gr/Default.aspx?tabid=389&sni%5B524%5D=1561&language=el-GR (in Greek).

[2] Op.cit 10.

– The concept and implementation of criminal liability is reinforced and parallel to the Directive for environmental liability this framework promotes the concept of the "polluter pays" but also the principles of prevention and precaution for a more effective protection of the environment and public health. This action is also translated to the direct adoption of measures to prevent and remedy environmental damage, the cost of which is transferred to the polluters.

3 Quality of Surface and Groundwater in Asopos River Basin

3.1 Surface Water

Regarding the quality of surface waters according to the monitoring data (MoEPPW 2006, p. 81): "nitrates and phosphorus concentrations in Asopos River were high whilst BOD and D.O. measurements indicated pollution caused by high organic loads due to industrial and urban wastes as well as agricultural run offs in its catchment area". According to the same report it is also noted that concentrations of microorganics, priority substances (Decision No. 2455/2001/EC) and dangerous substances (Directive 76/464/EEC) at Asopos River were lower than the threshold values set for the water quality objectives by the National Legislation (Decision 2/1-2-2001).

In the context of a research program of the Hellenic Centre for Marine Research (HCMR) (Chatzinikolaou 2009) aiming to assess the ecological condition of the Greek rivers using biological parameters, water samples from Asopos were analyzed. The day of sampling was chosen so as to reflect the worst situation in terms of surface flow in the river while sampling took place in three points/sites.

From the below table (Table 2.1) it is concluded that the ecological quality of waters in the study area according to the measurements of the HCMR[3] is characterized according to the EU Directive 2000/60 as "bad" in the two stations ASOPOS_DW and THERMIDON and "poor" in the station ASOPOS_UP. In addition, according to the first results it was found out that in Asopos water lived only one larva (Largo) of a kind of fly, which can only live under conditions of zero oxygen.

Furthermore, following the research of the National Technical University of Athens led by Professor Loizidou for the estimation of the environmental degradation because of the discharge of industrial outflows in the environment, sampling took place during June 2004, at different points of the area (along the river and at the coastal area of the estuary). The results showed that waters along the river and at the coastal area are polluted with inorganic and organic load.

[3] http://www.hcmr.gr/en/

Table 2.1 Results of sampling 2/8/2008[a]

			D.O. (%)	D.O. (mg/l)	T (°C)	Conductivity (mS/cm)	TDS (mg/l)	Salinity (PPt)	pH	N-NH4 (mg/l)	N-NO2 (mg/l)	N-NO3 (mg/l)	P-PO4 (mg/l)
*Ορια Οδηγίας 98/93				5		2.5			> 6.5 and ≤9.5	0.5	0.5	50	
			>75 %*										
SiteName	*Date*	*Time*											
ASOPOS_UP	2/8/2008	12:30	91.0	7.31	26.3	0.382	181	0.1	8.39	0.012	0	0	0.43
THERMIDONAS	2/8/2008	13:15	1.5	0.10	26.3	2.900	1,456	1.5	7.73	0.433	0.016	0.3	21.6
ASOPOS_DW	2/8/2008	14:00	1.2	0.10	26.5	3.690	1,853	1.9	8.62	0.295	0	0.13	2.78

Source: Chatzinikolaou (2009)

[a]Second raw of the table presents the limits/standards set by the Directive 98/93, while the asterisk is referred to limits/standards of the previous Directive

*Ορια προηγούμενης Οδηγίας

3.2 Groundwater Quality

Based on the study of the Institute of Geology and Mineral Exploration (Gianoulopoulos 2008b: cited in Technical Chambers 2009, p. 22) the qualitative characteristics of the groundwater were examined and classified into two different categories: the ions and the trace elements. Regarding the distribution of the main ions it was concluded that: "(i) In the area between Avlona, Ag. Thoma and Asopos River there are fresh waters enriched in HCO_3^- with a low total of dissolved salts (TDS). Eastern and at both sides of Asopos river bed the concentrations of TDS gradually increase and their hydro chemical type advances to Mg-HCO3 and to mixed type of waters. (ii) The majority of the groundwater is overloaded with nitrates demonstrating concentrations almost double compared to the 50 mg/L legal limits for drinking water. In addition, their concentration in Cl^- ions is increased while along Asopos it was also observed increase in PO_4 ions." Furthermore, in specific areas there were recorded increased values of Nitrite and Ammoniacal ions.

Following the same study of the Institute of Geology and Mineral Exploration (Gianoulopoulos 2008b: cited in Technical Chambers 2009, p. 22): "(i) The source of the nitrates is mainly agrochemical (nitrogen fertilizers) while the presence of nitrite and Ammoniacal ions is due to the urban and industrial pollution sources, while (ii) the increased values of Cl^- and PO_4 ions are also due to the industrial pollution sources. The increase of the Cl^- ions in the alluvial coastal aquifers of Oropos and Avlida is due to the mechanism of salinization."

Finally, an alarming groundwater related quality parameter which made its presence in the area is that of hexavalent chromium Cr(VI). Economou-Eliopoulos et al. (2011, p. 40) note that: "although the control of the chromium content in the industrial wastes is a requirement the Assopos basin (Thiva, Oinofita, Oropos, Avlona) was known from the impact of groundwater by Cr(VI), due to intensive industrial activity at this area, such as plating metals, leather processing, stainless steel, Cr-bearing alloys resistant to corrosion and oxidation, which may be discharged into the environment." As shown in Fig. 2.3 and commented by the authors (p. 39): "Groundwater samples from the Asopos aquifer showed a wide spatial variability, ranging from <2 to 180 ppb Cr_{total} content [almost same to the Cr(VI)-values] despite their spatial association. The presence of Cr(VI)-contaminated groundwater at depths >200 m is attributed to a direct injection of Cr(VI)-rich industrial wastes at depth rather than that Cr(VI) is derived from the Asopos river or by the interaction between water and Cr-bearing rocks".

In Vasilatos et al. (2008) study it is documented that (pp. 61–62): "Concentrations over the maximum acceptable level for Cr_{total} in drinking water (50 lg/L), according to the EU Directive (EC 1998), were found in several groundwater samples from Thiva – Tanagra – Malakasa basin. The contamination of groundwater by Cr(VI) that was found in the majority of groundwater samples in the Thiva – Tanagra – Malakasa basin has been related to the widespread industrial activity for the last 40 years and the usage of hexavalent chromium in various processes. … In the Asopos River, that crosses the Thiva – Tanagra– Malakasa basin, although total chromium

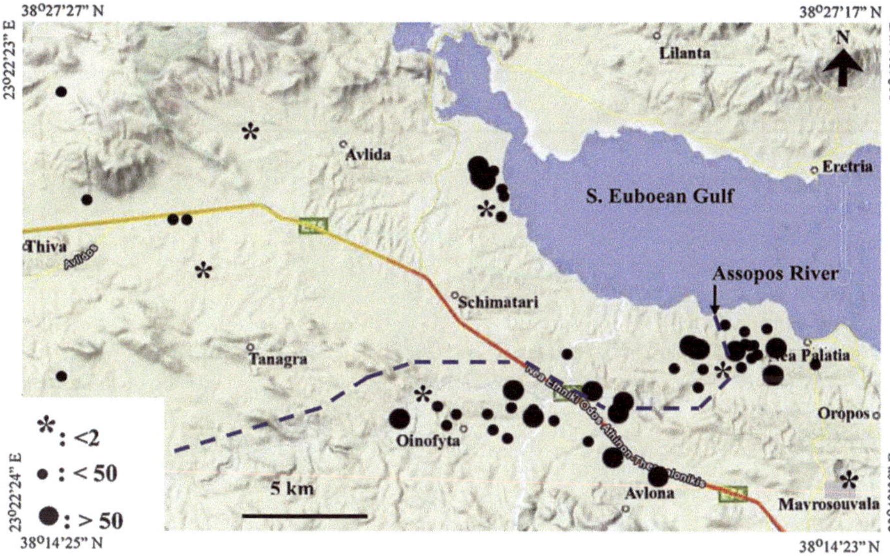

Fig. 2.3 Location of the groundwater boreholes and distribution of the Cr_{total} concentration (in ppb) throughout the Asopos basin. The groundwater samples were classified into three groups: (**a**) lower than 2 ppb Cr_{total}, (**b**) lower than 50 ppb Cr_{total}, and (**c**) higher than 50 ppb Cr_{total} (Source: Economou-Eliopoulos et al. (2011, p. 45))

is up to 13 Ig/L, hexavalent chromium less than 5 mg/L and the other toxic element concentrations were relatively low during our research, their values suggest their relation with the industrial activity in the area."

"The higher values of Cr_{total} are observed in the area of Avlida (180 ppb) and in the areas of both sides of Asopos river bed (40–107 ppb) till the broader area of Oropos (17–85 ppb), as well as in the area Southeast of Asopos till the area of Avlona (20–118 ppb). Increased concentrations were also observed in a borehole in proximity to Asopos river bed at North of Ag Thomas (163 ppb)" (Technical Champers, p. 23). As commented in the Technical Champers (2009) report in most of the cases these increased values of Cr_{total} are the result of industrial pollution, while the concentrations of Cr(VI)/(Cr_{6+}) demonstrate a similar distribution to that of Cr_{total}.

4 Pressures and Impacts of Pollution in Asopos River Basin

As it must be obvious to this point the broad Asopos area is the largest industrial region of Greece, supporting 1,300 industrial facilities such as metal processing agrochemical, and food/drinks industries among others. Hence, the industrial activity in the area of Asopos, which started at the end of 1970, and in particular in

Table 2.2 Industrial units distribution and average daily flow of wastewater for each industrial activity (m^3/day)

Industrial sector	# of industrial units	Average daily flow of wastewater (m^3/day)
Livestock farms	11	539
Food and drink industries	41	2,198
Textile and leather industries	20	1,925
Woodworking industries	15	2
Pulp and paper industries and printing industries	17	8
Chemical industries	77	154
Non metallic mineral industries	20	908
Metallurgy related industries	96	1,615
Commercial industries	9	–
Warehouses	26	–
Other industries	46	256
Total	**378**	
Closed	**28**	
Total	**350**	**7,605**

Source: Loizidou (2009)

Oinofita-Schimatari is considerable and increasing. According to the results of Loizidou (2009) study presented in Table 2.2, there were reported 378 industrial units in the area of which 23 were closed and the rest 5 were new and under construction. From the total of the existing units the 130 produce wastewaters during their production function. The total daily produced quantity of wastewaters reaches the 9,044 m^3/day of which 84 % corresponds to the industrial wastewaters. More particularly, the total daily produced quantity of industrial wastewaters from the total of the industrial and artisanal units located in the area reaches the 7,605 m^3/day and the total daily quantity of wastewaters of the employed in these units reaches the 1,439 m^3/day. The main flow of the produced industrial wastewaters is coming primarily from the sectors of "Textile and leather industries", "Metallurgy related industries" and "Industries of Foods and Drinks" at 25 %, 21 % and 30 % respectively.

Loizidou (2009) also offers a disaggregated view of the above industrial activities for each active industrial unit in Asopos RB as far as wastewater in m^3/day is concerned. Table 2.3 presents the industrial pollutants (Kg/d) in the current situation (during the research).

Loizidou's (2009) study explores also the potential for the creation of a Central Wastewater Treatment Plant (CWTP) and of a Central Sewerage Network of the Industries (CSNI). For this purpose it is important to predict the quantity and quality of the produced wastewaters in the future. Results are presented in the following table giving us an indication of the trend of the industrial development in the area. Hence, apart from the Textile and Leather industries all the other sectors demonstrate an increasing activity. Furthermore, in order to assure applicability of the

Table 2.3 Industrial pollutants (Kg/d)

BOD	203.8
COD	6,279
SS	2,275
Fats (Λίπη;)	765.5
MBAS	12
NH_4^+	
NO_3^-	46.95
PO_4^{-3}	13.86
SO_4^{-2}	2.774
CN^-	0.25
Phenols	16.66
AI	3.52
Fe	5.38
Sn	
Cr^{+3}	4.3
Cr^{+6}	0.00
Cu	2.94
Cd	0.35
Pb	1.71
Ni	42.27
Zn	2.11

Source: Loizidou (2009)

Table 2.4 Average daily flow of wastewater for each industrial activity (m^3/day)

Industrial activity	Current	10 years	20 years	20 years +15%
Livestock farms	539	807.9	1,091	1,254
Food and drink industries	2,198	3,298	4,451	5,119
Textile and leather industries	1,925	1,925	1,925	2,214
Woodworking industries	2	2.9	3.8	4.4
Pulp and paper industries and printing industries	8	12.2	16.4	19
Chemical industries	154	231.2	312.1	359
Non metallic mineral industries	908	1,362	1,839	2114.9
Metallurgy related industries	1,615	2422.8	3269.8	3760.3
Other industries	256	384	518.4	596.2
Total	7,605	10,445	13,427	15,441

Source: Loizidou (2009)

project even when faced with a higher volume of wastewaters than those predicted a 15 % (last column of Table 2.4) safety factor is added to the calculation of the future loads.

Regarding water consumption Table 2.5 shows that the most water demanding sectors are those of "Metallurgy related industries", "Food and drinks industries", "Textile and Leather industries" and "Chemical industries".

Table 2.5 Water consumption per industrial activity

Industrial activity	Quantity (m^3/year)
Food and drink industries	1,235,336
Livestock farms	35,650
Textile and leather industries	562,228
Woodworking industries	3,480
Pulp and paper industries and printing industries	4,015
Chemical industries	111,260
Non metallic mineral industries	40,805
Metallurgy related industries	1,240,540
Other	2,425
Total	3,235,739

Source: Loizidou (2009)

Table 2.6 Current ways of disposal of wastewaters in the study area

Ways of disposal	# of industrial units	Quantity of wastewater (m^3/day)	Percentage (%)
Surface disposal	15	454	6
Underground disposal	25	104	1.4
Recycling	22	946	12.4
Temporal collection	2	1	0.01
Municipal wastewater plant	19	95	1.3
Authorized management body	23	19	0.3
Asopos and its tributaries	24	5,985	78.7
Total	130	7,605	100

Source: Loizidou (2009)

As it is also noted in Loizidou (2009) from the total of the industrial and artisanal units that produce wastewaters, that is from the 130 units (37 % of the total of the units in operation) the 65 units (that is 50 % of these that produce waste water) have a treatment facility for their produced wastewaters. In general, from the total quantity of produced wastewaters (industrial and urban wastewaters) the 97 % is subject to a treatment. As Table 2.6 shows the existing ways of wastewater disposal in the area are the surface disposal, the underground disposal, the recycling of the treated out flow (within production activity), the disposal in a municipal wastewater treatment plant or in an authorized management body and the disposal in the Asopos River or its tributaries. Finally, some of the industrial units not having an alternative option collect temporally their wastes which are then managed by an authorized body.

Overall, Loizidou's (2009) study reports that considering the 350 industrial units that operate in the area the produced: a daily quantity of industrial wastewaters of 7605.0 m^3/day, 1438.8 m^3/day of sewage, 3458.034 m^3/year total annual water consumption and an annual quantity of sludge of 15,096 tn/year.

In addition, considering the study of the Agricultural University of Athens which recorded the industrial and artisanal units in the area it was concluded (Masoura, 2008: cited in Technical Champers 2009, p. 29) that: (i) North East of the Asopos' river bed in a stretch of 7 km there is a big number of industrial and artisanal units (407) which produce a daily volume of wastewaters ranging from 0.3 m^3/day to 3,000 m^3/day depending on the production procedure that is followed and the size of each industry, and (ii) these wastewaters are mainly produced by the textiles-varnishes-finishing, the food industries, the metallurgic units, the chemical units as well as from the livestock units.

An additional pressure in the area is also put by the agricultural activity which includes arable and tree crops. As stated in the Report of the Technical Champers (2009, p. 28): "there are 351,400 str. which are cultivated from which the 45 % are irrigated and the 55 % non-irrigated. In general, in the area an intense agricultural activity is observed in the upstream of Asopos." As it has been demonstrated in the previous section the production of the area is mainly characterized by hard wheat, pulses, cotton and some categories of fodder and vegetables. Therefore, the pollution of Asopos surface water from agricultural run offs is expected to aggravate the problem. Furthermore, in the same report it is noted that there are also natural sources of pollution apart from the industrial, urban and agricultural activities.

Apart from the obvious environmental impacts in the area there are also worrying human health impacts due to the presence of the highly toxic heavy metal hexavalent chromium (CrVI) (or chromium 6) which was traced at high concentrations (ranging from 10 ppbs to 330 ppbs) in both surface and groundwater samples from the area. As commented also previously the evidence show that the presence of hexavalent chromium in surface and groundwater is clearly and indisputably linked to industrial contamination of soils and waters from illegal discharge of industrial, hazardous waste and sludge. It is regarded that water consumption of high concentrations of hexavalent chromium harms human organism not only by drinking water but also by skin contact or the consumption of fruits and vegetables. Current scientific evidence has shown that hexavalent chromium is absorbed through the gastrointestinal tract and can reach many organs causing serious damages and cancer. Even the inhalation of hexavalent chromium can cause mainly lung cancer.

Furthermore, it has been noted that apart from the intense, human induced pollution of waters of Asopos River, which comes from the industrial activities in its catchment, in the coastal wetland complex the human influence is intense as well mainly due to expansion of settlements, impacts from traffic during the summer months, and garbage. A number of roads form blocks and the residential pressure goes up to the coastline. Indeed, the remnants of natural vegetation in some of the residential blocks indicate that parts of salt meadows have been filled to cross roads. Within the wetland toward the Oropos area the pressure from the filling and movement of vehicles is obvious. Part of the area at the mouth of Asopos River has been covered by the aviation with a variety of antennas and some of the area on both sides of the riverbed has been fenced.

5 Functions and Values of Asopos River and Oropos Lagoon

This section offers a short overview of the functions and values of the area, based on literature review and field visits, associated with the wetland complex composed from Asopos River (including its estuary) and the coastal lagoon of Oropos.[4] It is noted that Chap. 9 summarizes results of a qualitative evaluation of these wetland functions and related values by hydrogeomorphological unit.

The study area (Fig. 2.4) comprises the river and estuary of Asopos and the lagoon of Oropos northeast of Attica. Along the Asopos river human activities, mainly agriculture and industrial take place. For at least three decades industrial effluents pollute the river, the aquifers and the soils of the area, making it a negative example of the impacts to humans and environment that arise from non sustainable use of natural resources. The mouth of Asopos River and the lagoon of Oropos (on right hand side of Fig. 2.4) are located at around 2 km distance and are remnants of a single wetland system in the past associated with coastal marshes. Today, settlements have expanded in the in-between area. The road is located just few meters from the coast and halophytic vegetation is only sporadically present, resulting in the interruption of the structural continuity of the two wetlands.

Values for humans result from the structural features and functions of wetlands. These are the goods and services provided or potentially provided to humans as a result of functions that take place there. A wetland function can benefit humans in many ways, that is, to provide more than one values. Wetlands differ in the number of functions performed and in the degree to which they perform them; as a result, the values obtained by humans differ as well. Furthermore, many wetland values are derived from a combination of functions and a value may depend on one or more other values.

The functions that were considered necessary to evaluate in the area were: (a) water storage, (b) food web support, (c) nutrients removal and transformation, (d) sediment and toxic trapping, (e) floodwater attenuation, (f) groundwater recharge, (g) shoreline stabilization. The next sub-section presents in more detail the above mentioned functions.

5.1 Food Web Support

Wetlands, and especially those with a variety of vegetation units, depth and current velocity, and considerable vegetation cover, host a variety of habitat types. In the wetland complex of the estuary of Asopos River and the Oropos lagoon, characteristic vegetation types of the coastal zone appear, that create a variety of habitats for species and mosaic in physiognomy of the area. The change in the water depth due to the microtopography from the coast to inland area, also due to the periodical and seasonal inundation favors the development of vegetation types and the presence of species. With regard to the vegetation types, in sub sites (mouth of Asopos River

[4] EKBY (2010). Available at: http://www.aueb.gr/users/koundouri/resees/uploads/envstudyen.pdf

Fig. 2.4 Satellite picture of Asopos river and mouth, of Oropos lagoon and the wider area (Date: August 2004, Source: GOOGLE EARTH)

and Oropos lagoon), Mediterranean halophilous scrubs and Mediterranean salt meadows develop with more than one plant communities.

In particular, the Mediterranean halophilous scrubs (*Arthrocnemetalia fructicosae*) consist of four plant communities: (a) with dominant *Sarcocornia fruticosae* which occupies the largest surface area of the scrubs, (b) with *Halcocnemum strobilaceum* as dominant species, (c) with *Arthrocnemum macrostachyum* as characteristic species and (d) with characteristic species the *Aeluropus littoralis*. The Mediterranean salt meadows (*Juncetalia maritimi*) appear with two plant communities. The one is formed by the species *Elytrigia elongata* (=*Elymus elongates*) or *E. Flaccidifolius and Juncus acutus* as dominant in salt and wet soils, behind the zone established by halophilous scrubs, whereas in the other, *Juncus maritimus* and *Arthrocnemum macrostachyum* dominate. Moreover, in the area, depressions between dunes with reeds and sedge appear, and annual vegetation along the coastline. It is noted that the shallow lagoon of Oropos is partially connected with Evoikos gulf through a narrow channel and is a priority habitat type of Community importance (code 1150) according to Directive 92/43/EEC.

With regard to the diversity of animal species, we consider birds as a group of species indicative of the quality of the function, as it was previously described. In the study area, in total 101 bird species have been observed (Tzali et al. 2009), being the largest number compared to other wetlands of Attica region; they comprise raptors, herons, waterfowl present during the winter (November to February), waders all year round, with their peak during migration in winter. Important is considered the presence of *Larus melanocephalus* that is endangered according to the Red Data Book, among others. It is noted that 19 bird species belong to Annex I to the Birds Directive, out of which in the area the following reproduce: *Charadrius alexandrinus*, *Himantopus himantopus*, *Sterna albifrons* and *Calandrella brachydactyla*. The large number of bird species during migration shows the importance of the area as a stopover (Tzali et al. 2009, WWF Hellas and Hellenic Ornithological Society 2008 and on site visits 2010). However, with regard to fish, it is noted that endangered freshwater species such as *Pelasgus marathonicus* (Vinciguerra 1921) and Telestes beoticus (Stephanidis 1939) have probably been extinct from Asopos River (Economidis and Chrysopolitou 2009a, b). As regards reptiles, only the species *Ablepharus kitaibelii* has been observed in an oliveyard near the river, *Chalcides ocellatus* is also referred to. Moreover, around a kilometer north east of Oropos lagoon, the snake *Malpolon monspessulanus* has been observed (Ioannidis unpublished data 2002) and so its presence in the study area is probable. As to the mammals, there are no published data; however, it is probable that certain human acclimatized species such *Rattus rattus*, whose presence has been recorded in the study area, are present. It is also probable that some bat species also use the wetlands for food. In any case, the habitats for most reptiles and mammals are degraded. It is also noted, that the wetland complex, although small in size (around 100 ha), is the only remnant wetland between two larger coastal wetlands of the wider area, those of Sperheios and Shinias that belong to the NATURA 2000 network. As such, it contributes significantly to the conservation of habitats to the coherence of the network. This role is also acknowledged in article 10 of Directive 92/43/EOK, which encourages member states to conserve and manage those sites as special landscape features.

Given the above it becomes clear that the wetland complex performs the function Food web support. But the continuing deterioration of the function is obvious and results mainly from the expansion of settlements at the expense of wetland area and water pollution of the river Asopos. The upgrading of this function is directly related to the removal of these two threats. The expansion of settlements should not be pursued at the expense of wetland area and water quality of the Asopos must be restored so that the function is upgraded and maintained at a high degree. Furthermore, the development of targeted activities such as environmental information, education and recreation would probably prevent the degradation of salt meadows adjacent to the coast from vehicles especially during the summer months (the area is not flooded) and would encourage the sustainable use of the wetland complex.

5.2 Water Storage

The storage of surface water is a function that almost all the wetlands can perform. The storage of water from precipitation and the time required for the gradual discharge of the water varies between wetlands. The depth of water in a wetland provides a clear picture of the volume of water that can be stored. The wetlands that have shallow water warm faster than the deeper ones and have greater water losses due to evaporation. In contrast, wetlands with steep slopes maintain deep water throughout the area and have less water loss by evaporation. In the case of the Asopos River, despite the relatively large size of the basin, water is kept only for a minimum time in the river bed, due to rapid percolation to karstic aquifers in the basin. It therefore appears that the ability to perform this function is limited along the Asopos River but also in its estuary. Similarly limited is the extent to which this function is performed in the lagoon of Oropos. Limiting factors is the shallowness, the existing channel of water and the little inclined regions upstream of the lagoon, which contribute only little with water runoff.

5.3 Nutrients Removal and Transformation

The water entering a wetland inevitably carries nutrients dissolved in water and adsorbed to soil particles. The nutrients can come from the fields, villages and industries. Wetlands have the ability to improve water quality, through the retention of nutrients and their conversion into other forms. The removal is achieved by physico-chemical and biological processes that are performed in wetlands. The presence of plants favors the function of removal and transformation of nutrients, because they contribute to both slowing the water flow and absorbing the entering nitrogen and phosphorus. In addition, plants are the substrate for the growth of microorganisms, by which many substances are transformed into other forms. The wetland vegetation is particularly effective in the retention of nutrients. Therefore, the higher the coverage with vegetation, the more that function is favored. Moreover, this function is favored not only when there is vegetation in the wetland, but rather when there is a diversity of species. However, as mentioned above, Asopos River maintains permanent flow only for a limited time. Thus, this function is severely constrained by lack of water in the riverbed. In the lower reaches, where large quantities of liquid waste are disposed of, the presence of water is almost permanent, and despite the existence of wetland vegetation (especially reeds), nutrient concentrations are very high. Therefore this function seems to be limited along the riverbed and is further degraded in the section towards the exit to the sea. In terms of the lagoon of Oropos, theoretically there is the ability to perform this function. However, this is limited by the salinity of the water, which does not favor this function in general and the more specifically the presence of those plant species that support some of the processes of removal and transformation of nitrogen and phosphorus. The presence of plants as they grow around the lagoon, would enhance the degree of this function.

5.4 Sediment and Toxic Substances Trapping

The water runoff, atmospheric precipitation and wind drift within the wetland solid materials. These materials are trapped temporarily or permanently in the wetland so that the wetland improves the quality of water flowing through it. The sediments may contain nutrients, heavy metals and pesticides according to the land use of the catchment (agricultural, urban, industrial, etc.). The processes that trap pollutants entering into the wetlands, are due to slow speed of water and subsequent retention of nutrients by wetland vegetation, to the decomposition of organic matter by microorganisms, to the metabolism of plants and animals, and to photosynthesis. In the wetland of Asopos, the velocity of water is small and therefore there is a great likelihood of precipitation of particulate material, with subsequent retention and deposition of sediments and toxic substances within the wetland. The areas with wetland vegetation (especially reeds) reduce the speed of water in the wetland due to friction, contributing to the rapping of sediments and preventing re-suspension of sediments. Furthermore, the vegetation contributes to coating of pollutants and sediments with dead plant material, which is good feature. Also, the shallow water, contributes to the entrapment of pollutants and toxic substances, because the speed of the water is low due to friction and facilitates the trapping. The type of substrate determines the degree of adsorption of contaminants from the wetland soil. The presence of organic substrates favors the absorption and retention of certain toxic heavy metals and synthetic organic compounds. Therefore, wetlands with a high proportion of organic soils perform this function.

On this basis it is assumed that there is expression of that function, at the mouth of Asopos.

The lagoon can act as trap of sediment and toxic substances through sedimentation processes and coating with other materials or through biochemical processes in the water flowing in the lagoon from runoff from the nearby upstream region. The sediments sometimes carry, by chemical and physical bonds, toxic substances like heavy metals and pesticides. The site has features that enable it to perform this function to an appreciable extent; however, in practice, this function is limited because only limited amounts of water runoff flow through the Oropos lagoon.

5.5 Floodwater Attenuation

The function of amending the flood is very important for humans, due to the slowing of water velocity and the reduction of the area affected by flood. Wetlands can temporarily store large amounts of water during floods and mitigate the damage to agriculture and settlements. The ability of the wetland is to store flood water and gradually recharge, after the flood, thereby reducing the flood peak. This function is performed at an optimal level in ecosystems found in higher parts of the catchment, where the floods occur, and in wetland areas that are a recipient of torrential flow. The factors on which the efficiency of the function depends are: (a) the type of wetland and the morphology of the catchment, (b) the location and size of the catchment, (c) a channel of water, (d) water movement and (e) the vegetation cover.

In the case of Asopos River this function is not considered as particularly important given the high degree of infiltration of rainwater in the catchment through the carbonate rocks that prevail in the basin in approximately 50 %. It is most likely that underground water from side enters Asopos River which is moving slowly and not flood peaks from surface water. In the case of the lagoon of Oropos, this function seems to be performed quite well, protecting the residential area upstream from the sea water during extreme weather and tide, however limited it is. This is achieved through its ability to store and halt the rising waters and recharging them again to the sea through the channel.

5.6 Groundwater Recharge

Wetlands retain large amounts of surface water from the catchment area and they play a valuable role in the enrichment of underground aquifers, as water retained then flows through the soil layers to the aquifer, enriching it with water that would be lost otherwise. The enrichment is done both in the vertical and the horizontal movements of water, especially in cases of shallow aquifers. Whether a wetland has the potential to enrich the groundwater aquifers is mainly dependent on the substrate of the wetland that is the main factor in the performance of that function. If the substrate is practically impermeable, this function is not existent, even if all other factors are favorable. In the area of the Asopos River, there are two main aquifers; the upper aquifer of granular deposits and underlying karstic aquifer formations. The movement of groundwater varies vertically and horizontally between different aquifers and although the bed of the river in only small parts comes in contact with the karst aquifer, the river Asopos has no permanent water supply. Thus, it appears that a large volume of runoff water percolates to the groundwater and therefore the function of the enrichment of groundwater aquifers is performed widely.

For the Oropos lagoon no analytical data on the permeability of the substrate exist. From hydro-lithological data of the wider region it comes that formations of medium to low permeability dominate. Taking into account also the proximity to the sea and the possible entry of underground salt water, it is estimated that the lagoon does not perform this function.

5.7 Shoreline Stabilization

The stabilization of coasts by wetlands is associated with the retention of soil on the coast (e.g., from roots of wetland plants) to prevent the erosive action of waves and currents. The existence of specialized wetland flora, tolerant of increased salinity in coastal areas, plays a special role in this direction with their roots. However, resistant to erosion are not only wetlands with vegetation but also other coastal wetlands that are flat or nearly flat with no vegetation. The lagoons are a typical case of wetland ecosystems that absorb the energy of waves and currents, preventing them to enter the land and erode it. The lagoon of Oropos is assumed to perform this function until today, although the surface has been limited by residential development. Stabilization of coast is also achieved through the deposition of river delta wetlands. The delta, first absorb the erosive action of waves and currents and also provide the fine material which spreads and/or is deposited subsequently by waves, usually in the neighboring coasts, providing in this way double protection on the latter. Thus the estuary of Asopos River is assumed to perform this function that needs to be maintained.

With regard to the values it is noted that a value originates from one or more functions, whereas a function could shape more than one values. Wetland values could be separate, complementary or even contradictory. Wetlands differ in the number and magnitude of values. In the case of study area, the following wetland values were evaluated: (a) biological, (b) scientific, (c) education, (d) recreational, (e) hunting, (f) improving of water quality, (g) protection against floods, and (h) protection against erosion.

Detailed performance of the hydrogeomorphological units with regard to values is reported in Chap. 9. However, it is noted that one of the principal values of wetlands is biological value. It is related to the number of plants and animals, which can be considered rare or protected and their presence is associated with wetlands. The maintenance of high biodiversity contributes to the enhancement of values, such as scientific and educational. This value is directly affected by the function of food web support and indirectly from other functions. The study area has a biological value, since habitat types of Annex I of Directive 92/43/EEC, of which the lagoon is a priority for protection, are encountered. At the same time, the area has value in terms of birds, while for the rest of the fauna it appears to be limited because of its conservation status and of the threats it faces. Furthermore, wetland biodiversity makes them attractive locations for research from many disciplines. They are living laboratories for the observation and therefore frequently used for research. The area has vegetation units and plant and animal species that require scientific research and study (e.g. birds, habitat types of Annex I of Directive 92/43/EEC). The biological value of the region, particularly in the lagoon of Oropos, combined with the characteristics of the surrounding area make it attractive from a scientific and educational aspect, but this value is expected to rise following the implementation of management measures.

Wetlands are also highly valuable in terms of recreation. They offer possibilities for viewing and enjoying the landscape, observing flora and fauna, sports, excursions, walks and general mental and physical exaltation and contact with nature. In the case of river Asopos this value is very limited, whereas in the lagoon of Oropos it is assumed that the area has recreational value, which, after application of management measures would increase. Finally, wetlands, through the function of trapping sediment and toxic substances and the transformation and removal of nutrients, have the capacity to improve the quality of water flowing through them. This value is very poor especially for Asopos River. The impact of this (social, economic, environmental), is already visible in the catchment of wetlands under study whereas intense are considered the pressures on the coastal ecosystem of the South Evian Gulf, because of the direct relationship and proximity to this.

6 Conclusions

Although the impact of the industrial activity on the water use is not that important compared to other uses (irrigation, water supply), it is observed that some of the industrial units have a considerable production of wastewater of diverse pollutants which depend on the type of production. Regarding the estimation of the pressures on the water quality due to the industrial activity, it is necessary to analyze in detail the total amount of polluting material in the area as well as the part of the polluting material that is related to the industrial activity. In general, it is observed that the sectors of Basic metals and Foods and Drinks which are of great importance in the area produce a considerable number of wastes compared to the other sectors of the industrial activity.

It should be noted that apart from the industrial wastewaters which are the byproduct of production, there are also urban wastewaters that come from the employees of the relative units. The fact that the basin is in close proximity to Athens has played an important role due to the special conditions of the capital the last years as a result of the population growth, the increase in land prices, the urbanization of suburban areas. These are some of the factors that made difficult the establishment of industrial units in close proximity to Athens and indicated the case study area to be a more attractive option.

As it will be also emphasized in Chaps. 3 and 9 the agricultural activity in the area is important for the local economy while at the same time puts a considerable pressure on water use. Regarding its effect on water quality it is estimated that it plays an important role and is mainly due to the nitrates run off due to the excessive use of fertilizers. However, the impact of the sector is even more important regarding water quantity. In order to estimate correctly this impact it is necessary to record the way of water supply for irrigation purposes as well as the existing boreholes.

As commented in Loizidou (2009) the considerable industrial activity combined with the lack of necessary infrastructure results in serious pollution problems in the wider area. In particular, the lack of a holistic plan for the treatment of the produced liquid, solid and air wastes creates important environmental problems. A main problem is the uncontrolled disposal treated or not industrial wastewaters in different natural receptor and mainly in Asopos River or its tributaries leading in the surcharge of the stream. It should be also noted that despite the fact that the majority of the existing in the area industrial units has a system of wastewater collection and treatment, the problem of the degradation of the water quality of the receptors and in general of the aquifer of the area remains severe. This fact is due to either the non satisfactory operation of the wastewater systems because of technical or operational problems or to the small degree of effectiveness of the existing wastewater treatment systems. All these factors have contributed to characterize Asopos as one of the most polluted rivers having an impact not only to the areas that it crosses but also to the coastal area of Chalkoutsi in which it flows into.

Biological, scientific, educational, recreational and other different wetland values were evaluated in the study area. The maintenance of high biodiversity contributes to the enhancement of such values. This value is directly affected by the function of food web support and indirectly from other functions. It should be stressed that wetland values could be separate, complementary or even contradictory.

Finally, as a solution to the problem Loizidou's (2009) study explores the possibility of the creation of a management system for wastewater treatment which will involve the collection, transfer and treatment of the produced wastewaters. It has to be noted that the optimal management of the wastewaters will contribute significantly to the restoration not only of the Asopos River but also to the environmental enhancement of the area and will decrease the negative environmental consequences leading to a sustainable development. In addition, measures are deemed important as a result of the serious impacts of pollution on human health as well as of the high significance of biological, scientific-educational and recreational values of the study area.

References

Apostolopoulos, K. (2010). *Assessment of the economic importance of water use in the industrial and agricultural areas of Asopos RB*. Athens University of Economics and Business, Departement of International and European Studies. Available at: http://www.aueb.gr/users/resees/en/econcharen.html (In Greek).

Chatzinikolaou, G. (2009). *Network development and monitoring of the quality of the surface, transitional and coastal waters of the country – Assessment/classification of their ecological status*, Hellenic Centre for Marine Research (HCMR).

Commission of the European Communities (CEC). (2000). *Directive of the European parliament and of the council establishing a framework for community action in the field of water policy*. 1997/0067 (COD), C5-0347/00.

Economidis, P. S. & Chrysopolitou, V. (2009a). Pelasgus marathonicus (Vinciguerra, 1921). Pages 113–115. In A. Legakis & P. Maragkou (Eds.). *The red book of threatened animals of Greece* (528p). Athens: Hellenic Zoological Society.

Economidis, P. S., & Chrysopolitou V. (2009b). Telestes beoticus (Stephanidis, 1939). Pages 122–123. In A. Legakis & P. Maragkou (Eds.). *The red book of threatened animals of Greece* (528p). Athens: Hellenic Zoological Society.

Economou-Eliopoulos, M., Megremi, I., & Vasilatos, C. (2011). Factors controlling the heterogeneous distribution of Cr(VI) in soil, plants and groundwater: Evidence from the Assopos basin, Greece. *Chemie Erde – Geochemistry, 71*(1), 39–52. doi:10.1016/j.chemer.2011.01.001.

EKBY. (2010). *Functions and values of Asopos River and Oropos Lagoon: A preliminary analysis*. Available at: http://www.aueb.gr/users/koundouri/resees/uploads/envstudyen.pdf

European Communities. (1998, December 5). Council Directive 98/83/EC of 3 November 1998 on the quality of water intended for human consumption. *Official Journal of the European Communities*, L 330. Brussels: European Commission. Available at: http://eur lex.europa.eu/LexUriServ/LexUriServ.do?uri=OJ:L:1998:330:0032:0054:EN:PDF

Hellenic Republic. Ministry of Environment Physical Planning and Public Works (MoEPPW). (2003). *The law for the management of water resources of Greece* (Law N.3199/03). Athens.

Hellenic Republic. Ministry of Environment Physical Planning and Public Works (MoEPPW). (2006). *Report on the pressures and qualitative characteristics of water bodies in the water districts of Greece and a methodological approach for further analysis*. Athens.

Hellenic Republic. Ministry of Environment Physical Planning and Public Works (MoEPPW)-National Technical University of Athens (NTUA), Department of Water Resources and

Environmental Engineering. (2008). *National Programme for the Management and Protection of Water Resources.*

Loizidou, M. (2009). *Integrated management of industrial wastes and waste waters of the broader area of Oinofita- Schimatari.* National Technical University of Athens, School of Chemical Engineering. Project assigned by the Hellenic Environmental Inspectorate (HEI) of the Ministry of Environment, Energy and Climate Change (MoEECC) (In Greek).

Technical Chamber of Greece. (2009). *The problem of Asopos River-Suggestions for facing it. Report*, July 2009. (In Greek)

Tzali, M., Fric, J., & Promponas N. (2009). *The birds of wetlands in Attiki. Bird Monitoring Programme on Wetlands of Attiki* (56p). Athens: Hellenic Ornithological Society.

Vasilatos, M. I., Economou-Eliopoulos, M., & Mitsis, I. (2008). Hexavalent chromium and other toxic elements in natural waters in the Thiva – Tanagra – Malakasa Basin, Greece. *Hellenic Journal of Geosciences, 43*, 57–66.

WWF, & Hellenic Ornithological Society. (2008) *Preliminary study of management, protection and enhancement of wetland Skala Oropou* (96p). Athens.

Chapter 3
The Economic Characterization of Asopos River Basin

P. Koundouri, N. Papandreou, M. Stithou, A. Mousoulides, Y. Anastasiou, M. Mousoulidou, A. Antypas, I. Souliotis, T. Mavrogiorgis, and K. Vasiliou

Abstract This chapter puts the emphasis on the economic characterization of Asopos River Basin (RB) in order to identify the members of the society that will bear the cost of improvements and will benefit from its implementation. Following the first step of Directive's implementation, requiring the economic characterization of the river basin, there is the need to evaluate the economic significance of water in the region, identify key economic drivers influencing pressures and water uses, examine how these economic drivers will evolve over time and will influence pressures and how will water demand and supply evolve over time. Hence, an important objective of this chapter is to identify the economic significance of water uses, focusing on Asopos experience, considering primarily industrial and agricultural uses since residential and tourist related uses are the focus of Chap. 4.

Keywords Asopos River Basin • Economic characterization of water • Water uses • Water supply and demand • Water pricing

P. Koundouri (✉) • M. Stithou • Y. Anastasiou • M. Mousoulidou
A. Antypas • I. Souliotis • T. Mavrogiorgis • K. Vasiliou
Athens University of Economics and Business, Athens, Greece
e-mail: pkoundouri@aueb.gr; stithoumavra@yahoo.com; yanastasiou@aueb.gr; mmousoulidou@aueb.gr; aantypas@unipi.gr; souliotis.io@gmail.com; tmavrogiorgis@aueb.gr; kvasiliou@aueb.gr

N. Papandreou
Andreas G. Papandreou Foundation, Athens, Greece
e-mail: tulayev@gmail.com

A. Mousoulides
Athens University of Economics and Business, Athens, Greece

Intercollege Larnaca, Larnaca, Cyprus
e-mail: mousoulides.a@adm.lar.intercol.edu

P. Koundouri and N.A. Papandreou (eds.), *Water Resources Management Sustaining Socio-Economic Welfare*, Global Issues in Water Policy 7,
DOI 10.1007/978-94-007-7636-4_3,

1 Introduction

The first step in the implementation of the Water Framework Directive (WFD) (CEC 2000) is the economic characterization of water in the region.[1] A comprehensive economic characterization of the water in the region requires first of all that the economic significance of water in the region is evaluated. This involves an assessment of the industrial, agricultural, residential and tourism water needs in the area. However, as the focus of the chapter is the two first sectors, analysis will include information for example on the number of industrial units, the industrial water consumption, the total cropped area, cropping patterns, gross production and income of the farming population.

As the first step of the application of the economic characterization includes the economic evaluation of water use in the RB, the evaluation of the economic importance of water use in the area is realized for each sector of the economic activity in order to define the sectors that put more pressure in water use (see MoEPPW 2006, 2008). Every sector of economic activity (primary, secondary, tertiary and households) has a different water use. As a result, for the agricultural-livestock sector water use concerns the irrigation of groundwater or surface water for purposes such as the cultivation of land or to rearing of livestock. For the industrial- artisanship sector, water use takes place for washing and coloring (textiles), steel production, cement production, oil processing, energy production etc. Regarding the tourism sector of economy and domestic sector (households) water use concerns the water supply for home use by the authorized supplier. Water use in Asopos RB is relatively prioritized as follows:

i. Water supply of households, touristic units and holiday home
ii. Irrigation of cultivated areas and livestock units
iii. Industrial water use

In this context, the key economic drivers influencing pressures and water uses need to be determined including (a) the general socio-economic indicators such as population growth, income, and employment; (b) the key sector policies that significantly influence water use (e.g., agricultural and environmental policies); (c) the development of planned investments likely to affect water availability; and (d) the implementation of future policies (environmental and other) that is likely to affect water use. These economic drivers will need to be accounted in a dynamic perspective, i.e., to determine how these are likely to evolve over time.

The final component of the economic characterization of water in a region is the application of appropriate methodologies to assess sector-specific water demand. This involves deriving the marginal value of water in consumption and production, the price and income elasticity of demand, the marginal and average willingness to pay for public goods and quality changes of common access resources, and the associated risk parameters.

---

[1] Recall that the WFD principles are now incorporated in the law for water management resources of Greece (MoEPPW 2003).

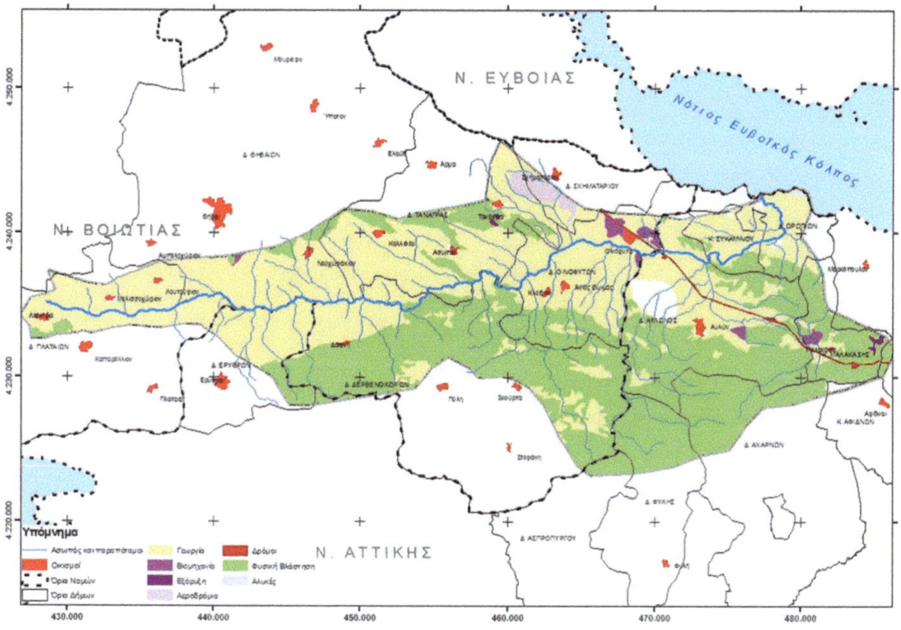

Fig. 3.1 Land use in Asopos RB (Source: Apostolopoulos 2010)

An overview of land use in Asopos RB is presented in Fig. 3.1 with the beige and light purple colors to represent agricultural and industrial use respectively. The use of specific indexes related to the above sectors of economic activity provides an indication of the economic importance of water use. Depending on the economic activity and consequently on the study area (industrial, touristic, residential, agricultural) the factors and indexes differ.

The parameters that are selected in order to be used as indexes of evaluation of the economic importance of water use are presented in the following sections.

2 Industrial Area of Asopos River Basin

Following Apostolopoulos (2010), the indexes that could be used to evaluate the economic importance of water supply in the industrial area of Asopos are:

i. The number of industrial units per sector
ii. Employment
iii. Total number of consumables
iv. Added value
v. Authorities and method of water supplying services
vi. Water pricing followed by water suppliers
vii. Water consumption

Table 3.1 Number of industrial units per sector in Asopos RB

Sector	Number of industrial units
Textile/dyeing/finishing	17
Metallurgy related industries	51
Food industries	21
Industries of detergents and cleaning products	6
Industries of production of chemical products	10
Industries of agricultural medicines and fertilizers	3
Dyeing industries	6
Pharmaceutical industries	4
Industries of plastic products	21
Industries of animal feed	3
Other industries	66
Chicken-livestock establishments	16
Total	224

Source: Technical Chamber of Greece (2009)

i. *The number of industrial units per sector*

The index describing the number of industrial units per sector in Table 3.1 provides information about the way the industrial sector is formed in the case study area. As a result, table shows that the most important sector is that of metallurgy related industries and then the sector of food industries, the sector of plastic products and the sector of textile/dyeing/finishing.

ii. *Employment*

Employment can be used for the estimation of the economic activity and the definition of the main sectors of the economy in the area. For that purpose Tables 3.2 and 3.3 report data on total employment as well as data on employment per industrial sector. According to the National Statistical Service of Greece[2] economically active are those who are employed and unemployed. The employed are classified as: (1) individuals of 10 years old and over who the week before the census worked even if that was for an hour and get paid or worked to help their family and (ii) individuals who have a job but the week before the census did not work due to illness, taking days off, because of a strike, weather conditions or for other reasons of temporal natural. Unemployed are those of 10 years old and over who stated that they are actively looking for a job. They are classified between: (1) individuals who lost their job for any kind of reason, and (2) young individuals who were looking for a job for a first time. Considering this terminology and the data reported in Table 3.3, the total economically active population (labor force) of Asopos RB was 18,243 in 2001.

[2] www.statistics.gr

Table 3.2 Economically active and inactive population in Asopos RB

Prefecture	Municipal department	Economically active population							Economically inactive
			Employed					Unemployed	
		Total	Total	Primary sector	Secondary sector	Tertiary sector	Did not state economic activity	Total	
Viotias	M.D. Dafnis	68	67	50	3	13	1	1	58
	M.D. Pylis	321	276	75	88	97	16	45	428
	M.D. Skourton	368	347	112	115	118	2	21	418
	M.D. Neochorakiou	292	272	241	6	22	3	20	200
	M.D. Agiou Thoma	728	685	248	256	151	30	43	588
	M.D. Kleidiou	144	130	42	57	29	2	14	193
	M.D. Oinofiton	3,079	2,834	158	1,212	922	542	245	2,460
	M.D. Leuktron	413	384	230	65	89	0	29	468
	M.D. Loutoufiou	153	147	93	15	34	5	6	129
	M.D. Melissochoriou	283	270	158	51	56	5	13	273
	M.D. Schimatariou	3,428	3,156	548	1,161	985	462	272	3,006
	M.D. Asopias	590	574	323	111	133	7	16	489
	M.D. Kallitheas	409	396	279	36	78	3	13	232
	M.D. Tanagras	562	501	178	131	142	50	61	507
Attikis	M.D. Avlonas	2,068	1,833	379	800	614	40	235	2,649
	M.D. Erithron	1,178	1,073	194	243	586	50	105	1,650
	M.D. Neon Palation	1,220	1,123	119	329	633	42	97	1,754
	M.D. Skalas Oropou	1,300	1,181	143	404	594	40	119	1,508
	M.D. Oropou	554	474	126	113	192	43	80	569
	M.D. Malakasis	564	522	31	167	313	11	42	750
	M.D. Sykaminou	521	480	90	144	233	13	41	684
Total (%)	**Asopos RB**	**18,243**	**16,725**	**3,817**	**5,507**	**6,034**	**1,367**	**1,518**	**19,013**

Source: National Statistical Service of Greece (2001 Census)

Table 3.3 Employees per sector in the Prefecture of Viotia

Sector	Employees in the Prefecture of Viotia	Employees in the whole country	Percentage (%)
Food and drinks	1,741	50,935	3.42
Tobacco	0	1,211	0.00
Textile materials	582	10,971	5.30
Leather and leather products	0	2,446	0.00
Woodwork and cork	0	2,192	0.00
Pulp and paper	278	5,332	5.21
Publishing and printing	0	13,371	0.00
Chemical products	888	14,664	6.06
Products of rubber and plastic materials	1,698	8,460	20.07
Products of non metallic minerals	1,078	16,012	6.73
Basic metals	3,126	9,865	31.69
Metallic products excluding machinery	1,291	12,843	10.05
Machinery and equipment	308	10,124	3.04
Electric machines and appliances	517	3,525	14.67
Other transport equipment	3,039	10,713	28.37
Furniture and other industries	184	2,393	7.69
Total	14,730	175,057	8.4

Source: National Statistical Service of Greece (personal contact)

Regarding 2001 census data 16,725 are employed (91.7 % of the labor force) and 1,518 are unemployed (8.3 % of the labor force). It is worth noting that for the whole country the respective rate of unemployment was 9.1 % in September 2009 while for the Prefecture of Sterea Ellada it was 9.7 %. The secondary sector is proved to be very important for Asopos RB since 32.9 % of the population is occupied in this sector. Furthermore, considering the fact that the total number of employed in the secondary sector in the Prefecture of Viotia is 13,506 (according to 2001 census), the 25 % of these are employed in Asopos RB demonstrating therefore a high industrial concentration in the area.

In order to analyze further the industrial activity, data per sector are reported in Table 3.3. The lowest level of analysis offered by the National Statistical Service of Greece is the Prefecture, however considering the concentration of the industrial activity in Asopos RB the main conclusions that are drawn by observing Viotia Prefecture can be considered to represent satisfactorily the case study area. On the other hand, regarding Prefecture of Attiki the industrial activity within the study site area is only a small percentage of the total Prefecture. Particularly, employment in the secondary sector in the Asopos RB within Attiki Prefecture is the 0.6 % of the Prefecture. Table 3.3 shows that the sectors of Basic metals and Other transport equipment play the most important role in terms of employment not only for the area but for the whole country. Also important, although demonstrating a less percentage of employment, are the sectors of Food and Drinks, Products of rubber

Table 3.4 Total consumables per sector in the Prefecture of Viotia

Sector	Total consumables in the Prefecture of Viotia	Total consumables in the whole country	Percentage (%)
Food and drinks	393,248,050	4,205,316,543	9.35
Tobacco	0	93,664,815	0.00
Textile materials	28,758,237	578,699,431	4.97
Leather and leather products	0	78,967,760	0.00
Woodwork and cork	0	119,584,404	0.00
Pulp and paper	15,107,816	399,564,692	3.78
Publishing and printing	0	560,127,151	0.00
Chemical products	62,815,814	1,464,315,634	4.29
Products of rubber and plastic materials	136,605,144	639,968,089	21.35
Products of non metallic minerals	129,432,576	1,560,346,128	8.30
Basic metals	1,588,879,071	3,335,063,768	47.64
Metallic products excluding machinery	113,760,436	987,946,274	11.51
Machinery and equipment	25,034,560	551,207,972	4.54
Electric machines and appliances	239,936,678	392,671,818	61.10
Other transport equipment	79,771,718	289,636,135	27.54
Furniture and other industries	5,412,537	81,478,565	6.64
Total	2,381,648,534	15,338,559,179	15.53

Source: National Statistical Service of Greece (personal contact)

and plastic materials, Metallic products excluding machinery and Products of non metallic minerals. Finally, it is noted that the sector of Electric machines and appliances is of considerable importance in relation to the corresponding employment in the whole country.

iii. *Total number of consumables*

According to the National Statistical Service of Greece the terminology "Total number of consumables" include the value of first and aid materials, the value of packing consumable material, of fuels, of electric power and the rest of consumable materials during the specific year. This index can be used for the estimation of the contribution of each industrial sector to the other economic activities from which its inflows result from.

Table 3.4 shows that of particular importance are the sectors of Basic metals and Electric machines and appliances not only for the study site area but also for the whole country. Furthermore, within Asopos RB the sector of Food and drinks is quite important while considering the whole country the sectors of Other transport equipment, Products of rubber and plastic materials and Metallic products excluding machinery have a considerable contribution.

Table 3.5 Added value per sector in the Prefecture of Viotia

Sector	Added value in the Prefecture of Viotia	Added value in the whole country	Percentage (%)
Food and drinks	254,197,230	3,076,272,044	8.00
Tobacco	0	158,920,035	0.00
Textile materials	14,694,267	367,556,710	4.00
Leather and leather products	0	75,011,709	0.00
Woodwork and cork	0	58,908,378	0.00
Pulp and paper	10,167,978	255,273,253	3.98
Publishing and printing	0	732,108,592	0.00
Chemical products	50,508,496	948,708,901	5.32
Products of rubber and plastic materials	78,652,072	390,838,151	20.12
Products of non metallic minerals	125,734,697	1,233,588,438	10.19
Basic metals	339, 459,178	1,042,092,592	32.57
Metallic products excluding machinery	65,606,262	633,186,191	10.36
Machinery and equipment	13,484,094	380,555,411	3.54
Electric machines and appliances	61,073,554	153,838,838	39.70
Other transport equipment	139,255,582	581,846,913	23.93
Furniture and other industries	8,583,994	77,267,486	11.11
Total	1,161,417,404	10,165,973,642	11.42

Source: National Statistical Service of Greece (personal contact)

iv. *Added value*

According to the National Statistical Service of Greece Added value is defined as the difference between the value of the produced products during a specific period of time, the revenues from service providing to third parties and any other revenues and the total of consumables. This index can be used to estimate the importance of each industrial sector in producing income and in the economic growth. Table 3.5 shows that the most important sectors for the study site and whole country regarding income production are the sector of Basic metals and the sector of Electric machines and appliances. Of special importance within the study site is the sector of Food and drinks, while compared to the whole of the country the sectors of Other transport equipment and Products of rubber and plastic materials are quite important.

v. *Competent bodies and method of water supplying services*

The water supply and sewage services are considered in the case of Greece as a public service. Across the country there are 214 enterprises for water supply and sewerage. The relevant competent bodies fall into three categories (Safarikas et al. 2006; Tsagarakis et al. 2003):

(i) In Athens and Thessaloniki cities – owned companies (non profit making corporations) own and operate the treatment plants (for water supply and wastewater) but function as private enterprises with a 20-year concession contract. In Athens, the Company for Water Supply and Sewerage of the Capital (EYDAP) and in Thessaloniki, the Company for Water Supply and Sewerage of Thessaloniki (EYATH). They are inspected by the Ministry of Environment, Planning, and Public Works (MoEPPW) that approves the pricing policy. The EYDAP S.A and EYATH S.A both serve the 53 % of the total Greek population. EYDAP S.A was established by the Law 1068/80 and the Greek State was shareholder, while EYATH S.A was established based on the Law 2651/1998 and the only shareholder was the Greek State. The first provides waters services to about 4,000,000 residents having 1,831,520 connections and 8,078 km of water supply network and sewage services to about 3,300,000 residents having a 5,800 km network of sewerage. The former serves 1,700,000 residents supplying 250,000 m^3 per day using a 1,800 km of water supply network and 1,450,000 residents supplying 170,000 m^3 per day using a 1,600 km network of sewerage. The networks account for 45 % of the volume supplied while the population served by EYDAP and EYATH is estimated to be 53 % of the total population of Greece.

(ii) Cities with more than 10,000 inhabitants are managed by Municipal Enterprises for Water Supply and Sewerage (DEYA) operating as private companies, but owned by the municipalities (Law 1069/80). However, there are also cities with population less than 10,000 inhabitants in which DEYA have been established. There are about 210 DEYA around the country from which 177 are organized and represented by the Union of Municipal Enterprises for Water Supply and Sewerage and provide their services to 3,500,000 residents. In total the DEYAs of the country serve the 35 % of the population through 1,300,000 connections for water supply. The area of DEYA's jurisdiction is defined as the area of the corresponding Municipality. According to the Law 2539/97 "Ioannis Kapodistrias" new Municipalities were created having a population of more than 10,000 residents resulting to the obligation of establishing a DEYA, since according to the Law 1069/80, a Municipality with population of over 10,000 residents that doesn't create DEYA is not entitled to be subsidized by the Public Investment Program (DEYA 65 % and Program 35 %). This is the reason that the number of DEYA was double from 105 in 1997 to 210. DEYA have as an objective the water supply and sewerage services provision while being responsible for the water quality, the early response to water shortage, the maintenance of the water supply and sewerage network, the construction of water supply projects etc. Population served by DEYA is estimated to be 35 % of the total population of Greece.

(iii) In the rest of the areas (towns/municipalities with less than 10,000 residents) the competent bodies are the Municipalities. These Municipalities which are responsible directly for water supply and wastewater services are about 830 and account for only 12 % of the total population served.

Focusing in the area of Asopos RB the above three independent systems are met as well as combination of these.

vi. *Water pricing followed by water suppliers*

Pricing of water consumption is a very important since it defines the credibility and the quality of water supply as well as the possibility of developing new programs for the satisfaction of future, quantitative and qualitative needs. According to the national program of management and protection of water resources that was planned by the Ministry of Environment Physical Planning and Public Works (MoEPPW), it is foreseen the definition of the water price, for all uses, through the relationship of cost-benefit. According to Article 9 of WFD (Article 8 P.D.51/2007), water pricing should have been applied till the end of 2010 so as:

- To provide the necessary incentives to the users in order to use effectively water resources and consequently to contribute to the achievement of the relevant environmental targets (Article 4 P.D. 51/07) and
- To establish the contribution of water use (industrial, irrigation and water supply) to recover the cost of water services, according to the principle of "polluter pays" and the economic analysis as described in the Annex IV of P.D. 51/07.

As a result, water pricing can contribute to the achievement of two particular objectives:

- The adoption of a plan for a sustainable environmental economy by reducing excessive consumption and optimizing the use of natural resources so to assure a satisfactory level of life.
- The establishment of a homogeneous way of gathering and processing necessary data for the expense accounting of water use.

As commented also in Chap. 1 through water pricing it should be aimed to:

- The recovery of financial cost (Capital Cost, Operation Cost, Maintenance Cost, Administrative Cost and Cost of Agricultural Subsidies) through the fees/payments for water supply and sewerage,
- The charge for excessive consumption which is to treated as pollution,
- The recovery of the environmental cost (associated with the impact on environment and social welfare as a result of the reduction in environmental quality),
- The assurance of regular water supply for the financially sensitive social groups and
- The recovery of the cost of natural resources (which is related to the reduction of the available quantity of the resource due to overexploitation).

Every system of providing services of water supply and sewerage such as EYDAP S.A, EYATH S.A, DEYA and Municipalities, follow a different water pricing policy since they differ in terms of operation and financial data which are to be considered in the pricing of water supply.

In particular, EYDAP S.A and EYATH S.A as privatized state owned and non-governmental companies set the prices which should however be approved by the state and cover their cost of operation. The pricing that EYDAP S.A employs defines ten categories of water use and users and imposes successive/escalating or standard charge and a monthly fixed charge depending on 3 months consumption.

According to EYDAP S.A billing for the industry:

- It is foreseen a compulsory consumption of 100 m^3/month
- For consumption up to 1,000 m^3 the charge is 0.8381€/m^3
- For consumption over 1,000 m^3 the charge is 0.9866€/m^3
- A monthly fixed fee ranging from 4.37€ to 43.62€ depending on the diameter of the water metering appliance

In addition, the billing system of EYDAP S.A has four categories of users regarding sewer use who are charged as a percentage of water price. In particular, as far as industry is concerned the right of use of sewage system is defined at 75 % of the value of consumed water excluding the sectors of beverages, ice-factories, paper-factories, and the industries of artificial silk for which the percentage reaches the 35 % of the value of the consumed water. Finally, pricing includes charges as V.A.T of 9 % of the value of consumed water and 19 % for the rest of uses as well as a charge of 1 % of the consumed water as a contribution to the fund responsible for the personnel's insurance.

Every DEYA which operates as a private company sets the prices which have to be approved from the City Council and cover the operational cost. Hence, for the calculation of the operational cost and the implementation of water pricing the following factors are considered:

- Expenses and salaries related to the personnel,
- Expenses and salaries related to third parties,
- Subsidies of third parties,
- Taxes-duties,
- Other expenses,
- Rates of interest and related expenses,
- Depreciation of capital etc.

However, WFD changes completely the way of operation of DEYA, and in particular their water pricing policy through the need for full cost recovery of water supply and sewerage. As a result, DEYA's water pricing policy should count in, apart from the financial cost, the cost of natural resources and the environmental cost.

Considering that the main size of the industrial activity is located in the Municipalities of Oinofiton and Schimatariou the following table (Table 3.6) presents the water pricing policy that is followed for industrial uses in these particular Municipalities.

It should be noted that many industrial units in the area of Oinofiton - Schimatariou are supplied with water directly from EYDAP S.A. and the corresponding pricing is used.

vii. *Water consumption*

This index includes the data of water consumption and the value of water according to the current pricing that the Municipalities apply. In the value of water the fixed fee is not included. The data regarding water consumption are coming from the Municipalities of Oinofiton and Schimatariou since the main size of industrial activity is located in these Municipalities (Table 3.7).

Table 3.6 Water pricing for industrial use in the Municipalities of Oinofiton and Schimatariou

Municipality	Frequency of bill	Fixed fee (€)	Pricing	Amount (€/m^3)
Schimatariou	3 months	85.28	Until 30 m^3	1.70
			31–90 m^3	1.84
			91–180 m^3	2.30
			181–270 m^3	3.05
			271–360 m^3	4.58
			More than 360 m^3	6.10
Oinofiton	6 months	14.67	Single	0.59

Source: Municipalities of Schimatariou and Oinofiton (personal contact)

Table 3.7 Water pricing for industrial use in the Municipalities of Oinofiton and Schimatariou

Municipality	Consumption (m^3)	Fixed fee (€)	Value of water (€)
Schimatariou	111,711	43,374	297,320
Oinofiton	240,788	2,963	142,065

Source: Municipalities of Schimatariou and Oinofiton (personal contact)

3 Agricultural Area of Asopos River Basin

The indexes that could be used for the assessment of the economic significance of water use for the agricultural area of Asopos RB are the following:

i. Agricultural land per crop
ii. Production per crop
iii. Yields per crop
iv. Irrigated land per crop
v. Subsidies
vi. Investments

Next each one of the indexes is analyzed as well as the way that each one can be used for the description of the agricultural activity in the area.

i. *Agricultural land per crop*

This index is consisted of the average land in "stremmata" (1,000 m^2) that was used for each crop during the years 2004, 2005 and 2006. This index can be used in order to describe the most important crops in the area (Table 3.8).

From the below table (Table 3.8) it is observed that cereals occupy the biggest part of the agricultural land in the study area (about 50 % of the total area) and specifically the hard wheat which was cultivated on average in 132,400 str. The tree crops follow (18 % of the total land) and specifically the olive trees for olive oil production which occupy on average 53,670 str. Industrial plants represent only the 8 % of the total land the same as vineyards while the vegetables the 6 % the same as forage. In total the main crops (hard wheat, olive trees, cotton, vineyards and dry onions) sum up to 80 % of the total agricultural land in the study site.

Table 3.8 Agricultural land per crop

Crop category	Agricultural land in Asopos RB (str.)	Agricultural land in the country (str.)	Percentage (%)
Cereals for grain	145,751	12,083,038	1.21
Tree crops	54,461	9,681,422	0.56
Industrial plants	24,482	3,744,148	0.65
Vineyards	23,311	817,967	2.85
Vegetables	18,894	989,177	1.91
Fodder plants	17,370	3,088,495	0.56
Potatoes	6,972	479,970	1.45
Melons/watermelons	6,097	259,804	2.35
Edible pulses	2,475	140,114	1.77
Total	299,814	31,284,133	0.96

Source: National Statistical Service of Greece, Annual Agricultural Statistical Survey 2004, 2005 and 2006 (http://www.statistics.gr/portal/page/portal/ESYE/PAGE-themes?p_param=A0403)

Table 3.9 Production per crop category

Crop category	Production in Asopos RB (kg)	Production in the country (kg)	Percentage (%)
Vegetables	71,887,800	2,941,680,656	2.44
Cereals for grain	31,298,950	4,767,737,027	0.66
Vineyards	25,559,133	807,575,239	3.16
Potatoes	16,605,333	965,423,636	1.72
Melons/watermelons	16,477,153	870,490,654	1.89
Tree crops	6,602,773	4,883,725,651	0.14
Industrial plants	6,262,383	1,182,360,583	0.53
Fodder plants	4,971,583	2,044,026,573	0.24
Edible pulses	425,250	30,454,670	1.40
Total	180,090,359	18,493,474,689	0.97

Source: National Statistical Service of Greece, Annual Agricultural Statistical Survey 2004, 2005 and 2006 (http://www.statistics.gr/portal/page/portal/ESYE/PAGE-themes?p_param=A0403)

ii. *Production per crop category*

This index consists of the average yield in kilos (kg) that was produced for each crop during 2004, 2005 and 2006. This index can be used in order to describe the most important crops that can be found in the area.

From the above table (Table 3.9) it is observed that vegetables are the 40 % of the total production and especially the dry onions which reported a 3 year production of 36,658,000 kg. Cereals for grain follow (17 % of the total area) and particularly the hard wheat with 28,567,817 kg of production and then the vineyards (14 % of the production). Of less importance is the potatoes production and melons/watermelons (about 9 % each) and of even less the tree crops, the industrial plants and the fodder plants. In total the main categories of crops (hard wheat, vineyards, dry onions, potatoes and melons/watermelons) sum up to about 70 % of the production in the study site.

Table 3.10 Production per crop category

Crop category	Yield in Asopos RB (kg/str)	Yield of non-mountainous areas of the country (kg/str)	Percentage (%)
Hard wheat	216	244	88
Non-irrigated cotton	124	187	66
Irrigated cotton	338	321	105
Dry onions	5,356	3,284	163
Olive trees	111	377	29
Vineyards, mainly for wine production	1,095	1,011	108
Vineyards, mainly for table grapes	1,844	1,759	105

Source: National Statistical Service of Greece, Annual Agricultural Statistical Survey 2004, 2005 and 2006 (http://www.statistics.gr/portal/page/portal/ESYE/PAGE-themes?p_param=A0403)

iii. *Yields per crop category*

This yield consists of the average yield per stremma (kg/1,000 m^2) for each of the main crops (hard wheat, irrigated cotton, non-irrigated cotton, olive trees) during 2004, 2005 and 2 006. This index compared to the corresponding prices for the total of the non-mountainous areas of the country, can be used to describe the quality of the soil and climate conditions of the area but also the intense of the agricultural practices. The choice of the non-mountainous areas is made for comparison reasons since the area of Asopos RB does not include mountainous regions as these are defined in the Directive 85/148/EE.[3]

From the above table (Table 3.10) it is concluded that the hard wheat, the vineyards and the irrigated cotton are comparable to the rest of the country, while non-irrigated cotton and mainly the olive trees fall short considerably. On the other hand, the dry onions present an increased yield in Asopos RB.

iv. *Irrigated land per crop*

This index includes the average irrigated land during the 3 years period 2004–2006 per crop category. It can be used for the estimation of the real needs in the area as far as irrigation is concerned. In addition, considering that the average price for water consumption for irrigation is €13.73 per stremma of irrigated land (Koundouri et al. 2009) it is possible to calculate the value of water for irrigation.

Observing the following table (Table 3.10) and considering the land per crop category, it is noticed that from about 203,000 str. which include plants of long-living plantations and other crops (cereals for grain, pulses, industrial plants, fodder plants, melons/watermelons and potatoes) only 28,444 str. were irrigated. From this area the higher percentage belongs to the irrigated cotton which occupies an area of about 15,000 str. The value of water for irrigation is estimated at €755,850 (Table 3.11).

[3]**EE** L 56 of 25.2.1985, p. 1 to 59.

Table 3.11 Irrigated land per crop category

Crop category	Irrigated land (str)	Cost of use (€)
Plants of long-living plantations	28,444	390,536
Vegetables	18,894	259,415
Tree plants	6,004	82,435
Vineyards	1,709	23,465
Total	55,051	755,850

Source: National Statistical Service of Greece, Annual Agricultural Statistical Survey 2004, 2005 and 2006 (http://www.statistics.gr/portal/page/portal/ESYE/PAGE-themes?p_param=A0403)

v. *Subsidies*

The subsidies that are provided to the produces in the framework of the first pillar of the Common Agricultural Policy are regarded that increase the cost of the resources. The resource cost in case of water represents the opportunity cost of water use compared to other present or future uses, which are excluded due to the current use of the resource beyond its degree of natural replenishment or recovery (European Communities 2002-WATECO). According to this definition subsidies could be considered that create an incentive for water use excluding it from current or future uses. For the right estimation of the cost of the resource it should be taken under consideration only that part of subsidies that can be correlated with water use. For that purpose water elasticity prices are considered in the production function of each crop category, which is estimated at 0.1 for cereals and pulses, 0.36 for industrial plants and 0.2 for olives (Koundouri et al. 2009). According to these prices and the total height of the subsidies the percentage of subsidies that corresponds to each water use can be calculated.

In the next table (Table 3.12) the total subsidies per Municipal Department in Asopos RB are presented as published by the Greek Payment Authority[4] of Common Agricultural Policy (CAP) Aid Schemes for 2008. Considering the fact that there is no division of the subsidies per crop category, the percentage of land that each crop occupies in each Municipal Department could be used so as to part the subsidies per crop category. For that purpose in Table 3.13 it is presented the percentage of land that each of the main crop categories occupies per Municipal Department in Asopos RB.

Combining both of the below tables (Tables 3.12 and 3.13) and considering the subsequent elasticities, Table 3.14 emerges in which the amount of subsidies that corresponds to water use is presented. The total amount reaches 818,799 € and is an estimation of the resource cost as a result of its current use in agricultural activities.

vi. *Investments*

This index includes the investments that took place in Asopos RB in the context of the "Operational Program for the Agricultural Development and Reform of the Country Side" for the period 2000–2006. The investments that are reported in

[4] Payment and Control Agency for Guidance and Guarantee Community Aid (OPEKEPE).

Table 3.12 Subsidies of the European Agricultural Guarantee Fund (EAGF) in Asopos RB

Prefecture	Municipality/commune	Municipal department	Direct Payments (€)
Viotias	M. Dervenochorion	M.D. Dafnis	440,019
		M.D. Pylis	557,395
		M.D. Skourton	592,804
	M. Thivaion	M.D. Neochorakiou	696,895
	M. Oinofiton	M.D. Agiou Thoma	966,363
		M.D. Kleidiou	349,450
		M.D. Oinofiton	121,402
	M. Plataion	M.D. Leuktron	789,778
		M.D. Loutoufiou	910,677
		M.D. Melissochoriou	775,253
	M. Schimatariou	M.D. Schimatariou	662,739
	M. Tanagras	M.D. Asopias	645,651
		M.D. Kallitheas	563,657
		M.D. Tanagras	617,985
Attikis	M. Avlonos	M.D. Avlonas	821,005
	M. Erithron	M.D. Erithron	1,855,814
	M. Oropion	M.D. Neon Palation	5,632
		M.D. Skalas Oropou	25,601
		M.D. Oropou	58,946
	M. Malakasas	M.D. Malakasis	36,293
	M. Sykaminou	M.D. Sykaminou	199,461
Total			11,692,819

Source: Payment and Control Agency for Guidance and Guarantee Community Aid (OPEKEPE). Data of year 2008. http://www.e-enisxyseis.gr

the area concern agricultural holdings (plans of improvement) and premium for the setting up in farming of young farmers, while there are no investments in the framework of the Program for agro-industry.

The relevant table (Table 3.15) shows that the biggest part of the investments in the context of the Program concerns the setting up in farming of young farmers and therefore it is expected an update in the age range of the agricultural population of the area as well as an improvement in the level of specialization of the occupied people in agriculture. This improvement in the human capital can lead to an increase of productivity of the agricultural sector of the area and can have both positive and negative impacts on the environment. The negative impacts can be the result of possible increased needs in inputs (fertilizers, water) because of the increase in the production, while the positive impacts can be ought to the higher awareness and specialization of the young farmers in issues of sustainable agriculture.

The investments in natural capital in agricultural holdings is more likely to lead to an increased need for inputs rather than to an reduction of inputs since these investments do not primary concern improvements in the methods of production towards a sustainable direction.

Table 3.13 Percentage of agricultural land per crop category in Asopos RB

Prefecture	Municipality/commune	Municipal department	Cereals & pulses (%)	Industrial plants (%)	Olive trees (%)
Viotias	M. Dervenochorion	M.D. Dafnis	92	6	2
		M.D. Pylis	60	3	37
		M.D. Skourton	99	0	1
	M. Thivaion	M.D. Neochorakiou	91	9	0
	M. Oinofiton	M.D. Agiou Thoma	59	7	34
		M.D. Kleidiou	73	1	26
		M.D. Oinofiton	54	2	44
	M. Plataion	M.D. Leuktron	48	15	37
		M.D. Loutoufiou	56	37	6
		M.D. Melissochoriou	48	39	13
	M. Schimatariou	M.D. Schimatariou	67	6	26
	M. Tanagras	M.D. Asopias	76	0	24
		M.D. Kallitheas	90	8	2
		M.D. Tanagras	42	17	40
Attikis	M. Avlonos	M.D. Avlonas	66	1	32
	M. Erithron	M.D. Erithron	74	24	2
	M. Oropion	M.D. Neon Palation	33	0	67
		M.D. Skalas Oropou	32	0	68
		M.D. Oropou	33	0	67
	M. Malakasas	M.D. Malakasis	10	0	90
	M. Sykaminou	M.D. Sykaminou	31	0	69

Source: National Statistical Service of Greece, Annual Agricultural Statistical Survey 2004, 2005 and 2006 (http://www.statistics.gr/portal/page/portal/ESYE/PAGE-themes?p_param=A0403)

4 Trends in Water Supply and Demand from Industry and Agriculture

Regarding the industry, in the case of the planning of an industrial zone in the area it is likely that the demand for industrial water use will increase considering the need of new land for the establishment of industrial units. As long as agriculture is concerned, during the period 2000–2006 there has been important investments especially for the reform of the agricultural population through the Operational Program for the "Agricultural Development and Reform of the Country Side" which are continued during the period 2007–2013. The population reform of the agricultural sector has as a result the increase of specialized farmers. Combining this effect with the objective of Common Agricultural Policy for a more sustainable agriculture is expected to increase the efficiency of water use and to limit the relevant demand.

In order for the monitoring of the demand to be feasible and to be able to apply measures for its potential limitation, the record and surveillance of all the existent bore holes in both the industrial and agricultural sector.

Table 3.14 Subsidies that correspond to water use in Asopos RB

Prefecture	Municipality/commune	Municipal department	Subsidies that correspond to water use (€)
Viotias	M. Dervenochorion	M.D. Dafnis	41,466
		M.D. Pylis	33,759
		M.D. Skourton	58,940
	M. Thivaion	M.D. Neochorakiou	65,456
	M. Oinofiton	M.D. Agiou Thoma	58,670
		M.D. Kleidiou	25,617
		M.D. Oinofiton	6,567
	M. Plataion	M.D. Leuktron	40,186
		M.D. Loutoufiou	58,158
		M.D. Melissochoriou	42,376
	M. Schimatariou	M.D. Schimatariou	45,587
	M. Tanagras	M.D. Asopias	48,802
		M.D. Kallitheas	52,026
		M.D. Tanagras	28,012
Attikis	M. Avlonos	M.D. Avlonas	54,684
	M. Erithron	M.D. Erithron	148,999
	M. Oropion	M.D. Neon Palation	184
		M.D. Skalas Oropou	814
		M.D. Oropou	1,942
	M. Malakasas	M.D. Malakasis	361
	M. Sykaminou	M.D. Sykaminou	6,193
Total			818,799

Table 3.15 Investments in Asopos River basin due to the "Operational Program for the Agricultural Development and Reform of the Country Side" for the period 2000–2006

Municipality/Commune	Investments in agricultural holdings (€)	Single premium for setting up a new holding (€)
M. Dervenochorion	8,252	489,400
M. Thivaion	760,818	1,885,100
M. Oinofiton	6,490	1,163,700
M. Plataion	332,110	1,883,100
M. Tanagras	386,593	1,904,100
Total	1,494,262	7,325,400

Source: Monitoring Information System (http://www.ops.gr/Ergorama/index.jsp?menuitemId=a1&tabid=0)

Regarding industrial water supply it is estimated that it will have a positive trend if the relevant announcements are applied, since the water quality that is provided to the industries will be improved. The agricultural sector does not seem to face a supply problem, since the water balance regarding the Sterea Ellada hydrometric area is positive. The competent bodies for providing water services of the Asopos RB are of particular interest. In the area there are one DEYA, eight Municipalities and one smaller administrative area identified as competent bodies for water provision. However, some of the above bodies provide their services through their collaboration with EYDAP. The way of water provision and treatment as well as the water abstraction points are diversified across the competent bodies and for that reason different quality, quantity and charges of water are observed. Due to the different way of operation of the bodies responsible for water provision (DEYA, Municipalities, and EYDAP) but also to the different financial elements that are included in the calculation of the height of the charge, the result is a highly diverse water policy within Asopos RB. This diversification is not only observed between the different systems of water provision (e.g., between DEYA and Municipalities) but it is equally intense within the same system of provision (e.g., across Municipalities).

5 Conclusions

This chapter offered an overview of the economic characterization in Asopos RB focusing on the sectors of industry and agriculture. In this context, specific indexes were studied in order to evaluate the economic significance of water use in the industrial and agricultural area of the basin.

According to the study of the Technical Chamber (2009) there are 244 industrial units in the area, which is 6 % of the total industrial units of the country. Therefore, it is observed a considerable concentration of industrial activity in the area, which is in need of a stable water supply. The water that is provided to the industrial units it is used either for the sanitation purposes of the employed or in the production procedure. The employees' needs are not negligible since in the industrial units of the area there 5,507 people. In this number there should be added a considerable number of employees who do not reside within Asopos RB but commute on a daily basis. Regarding employment the most important industrial sectors in the area are the sector of Basic metals and the sector of Other transfer equipment. Other important sectors are those of Foods and Drinks, Products of rubber and plastic material, Metallic products excluding machinery and Products of non metallic minerals. Finally, in the sector of Electric machines and appliances there are 517 people employed who although seem to be only few, they represent the 15 % of the total employment of the sector. The increased industrial activity in the area has considerable needs in raw materials and industrial goods consisting in that way an important consumer for other sectors of the economy. This is evident from the fact that in the area it is used the 15.5 % of the consumables.

In order to drive to conclusions regarding the importance of every sector as consumer of raw materials, it is necessary to compare the consumables of the area with the consumables of the whole country, since the raw materials and the industrial goods are not necessary to be produced in the area but to come from any other area of the country or even to be imported from other countries. As a result, the most important sectors are the sectors of Electrical machines and appliances and the sector of the Basic metals while considerable are also the sectors of Foods and Drinks,

Table 3.16 The importance of industrial sectors of Asopos RB for the whole country

Sector	Employment	Consumables	Value added	Total
Basic metals	+++	+++	+++	+++
Electric machines and appliances	+	+++	+++	+++
Other transfer equipment	+++	++	++	+++
Products of rubber and plastic materials	++	++	++	++
Metallic products excluding machinery	+	+	+	++
Foods and drinks	++	+		++

Other transfer equipment, Products of rubber and plastic materials and of Metallic products excluding machinery.

The industrial sector of the area produces also an important value added, which contributes considerably to the growth of the Greek economy. In particular, in the area it is produced the 11.5 % of the total value added of the corresponding industrial sectors of the country (about 1.2 billion Euros). Comparing the value added per sector with the whole of the country it is observed that the most important sectors are the sectors of the Basic metals and the sector of Electrical machines and appliances. Of considerable importance within the basin is the sector of Foods and Drinks, while in relation to the whole of the country important are the sectors of Other transfer equipment and Products of rubber and plastic materials.

From the above it is concluded that from the sectors that are active in the area, the most important are, compared to the whole country, the sectors of Basic metals, the sector of Electrical machines and appliances and the sector of Other transfer equipment (Table 3.16). Then the sectors of Products of rubber and plastic materials and Metallic products excluding machinery follow, while of less importance is the sector of Foods and Drinks. Especially for the sector of Foods and Drinks it is worth noting that while it seems to be of less importance compared the whole country, it is important compared to the rest of sectors of the area since this sector:

- Occupies the 12 % of employed in the industry of the area
- Consumes the 17 % of the total consumables of the area
- Produces the 22 % of the total value added of the area

Regarding the agricultural area within Asopos RB it is estimated that about 300,000 str. are cultivated which correspond to the 1 % of the total cultivated land of the country. The term "relative cultivated land/farmland" defines the total of land in the county in which crops corresponding to these of the study area are cultivated. The most important categories of crops in the area are the hard wheat, the olive trees for olive oil, the cotton, the vineyards and the dry onions while compared to the "relative cultivated land" of the whole the country the most important crops are the dry onion (14 % of the country), the non-irrigated cotton (9 % of the country), the melons and the vineyards for wine production (about 5.5 % of the country for each of them).

Considering the produced quantities, in the area the total of production is 180 million of kilos and it represents the 1 % of the relative production of the total of the country. The term "relative production" describes the total of the country's production that is produced by crop categories that correspond to these of the study area. The most important crops for the area are the hard wheat, the vineyards, the dry onions, the potatoes and the melons/watermelons while compared to the country's relative production the most important crops area the dry onions (23 % of the country), the non-irrigated cotton and the vineyards for wine production (6 % of the country for each crop) and the melons (5 % of the country).

From the above it is concluded that the most important crop categories are the hard wheat, the vineyards, the cotton, the olive trees for olive oil production, the dry onions, the potatoes and the melons. Especially the dry onion, the non-irrigated cotton and at a less extent the vineyards for wine production have a considerable contribution to the country's relative production. The importance of these crops can be attributed to the very satisfactory yields that these crops demonstrated in the area compared to the total of the country.

Regarding the irrigation, the average irrigated land during 2004–2006 was about 20 % of the cultivated land. A considerable part of it is covered from the irrigated cotton (about 25 % of the irrigated land), which as it has been demonstrated it does not consist an important crop for the area while its contribution to the total of the country is negligible. Another important part of irrigations is related to the vegetables (about 30 %) which seem to be important not only for the area but also for the whole country, while they are also related to the sector of Foods and Drinks, which is also of considerable importance in the area.

It should be also noted that in the area of Asopos RB important subsidies are provided. In 2008 the size of the subsidies reached about €11. 5 M distributed to about 4,162 beneficiaries. It is also noted that a considerable number of beneficiaries are not exclusively employed in the agricultural sector. The average number of subsidies reached the 2,800 € per person while the maximum amount reached even the 98,309 €. The subsidies could be related to the resource cost, considering that they may creative incentives for agricultural water use not allowing to be used for other present or future uses. Of course, the decoupling of the direct payments from the production reduces that potential while the perspectives of CAP foresee a support mechanism which is completed decoupled and which will correspond to the support of the income and will have as a prerequisite the compliance with the rules of cross compliance and a series of additional payments which will include among others payment for the environmental public goods which are produced by the agriculture through measures such as fallow land, crop rotation, permanent pastures and plant cover.

Overall, from this chapter it is concluded that both industry and agriculture in the Asopos RB play an important role not only for the area but also for the whole country. In this context, water as one of the most important inputs for both sectors consists a considerable factor of the area's economy.

References

Apostolopoulos, K. (2010). *Assessment of the economic importance of water use in the industrial and agricultural areas of Asopos RB*. Athens University of Economics and Business, Departement of International and European Studies. Available at http://www.aueb.gr/users/resees/en/econcharen.html (In Greek).

Commission of the European Communities (CEC). (2000). *Directive of the European Parliament and of the council establishing a framework for community action in the field of water policy*. 1997/0067 (COD), C5-0347/00.

European Communities 2002. *Economics and the environment – the implementation challenge of the Water Framework Directive. A guidance document* (Guidance document No. 1). Common Implementation Strategy for the Water Framework Directive (2000/60/EC). Brussels: European Commission. Available at http://forum.europa.eu.int/Public/irc/env/wfd/library.

Hellenic Republic. Ministry of Environment Physical Planning and Public Works (MoEPPW) (2003). *The law for the management of water resources of Greece*, Law N.3199/03. Athens, Greece.

Hellenic Republic. Ministry of Environment Physical Planning and Public Works (MoEPPW). (2006). *Report on the pressures and qualitative characteristics of water bodies in the water districts of Greece and a methodological approach for further analysis*.

Hellenic Republic. Ministry of Environment Physical Planning and Public Works (MoEPPW)-National Technical University of Athens (NTUA), Department of Water Resources and Environmental Engineering. (2008). *National programme for the management and protection of water resources*.

Koundouri, P., Remoundou, K., & Kountouris, Y. (2009). *A Note on the Implementation of the Economics of the EU Water Framework Directive under Data Limitations: A rapid appraisal approach*. DEOS Working Papers, Athens University of Economics and Business.

National Statistical Service of Greece, Annual Agricultural Statistical Survey 2004, 2005 and 2006.

Safarikas, N., Paranychianakis, N. V., Kotselidou, O., & Angelakis, A. N. (2006). Drinking water policy in the frame of the Directive 2000/60/EC with emphasis on drinking water prices. *Water Science and Technology: Water Supply, 5*(6), 243–250.

Technical Chamber of Greece. (2009, July). *The problem of Asopos River-Suggestions for facing it*. Report, (In Greek).

Tsagarakis, K. P., Paranychianakis, N. V. & Angelakis, A. N. (2003). Water supply and wastewater services in Greece. In S. Mohajeri et al. (Eds.), *European water management between regulation and competition, aqualibrium project, EU-Directorate-general for research, global change and ecosystems, B-1049*, Brussels, Belgium, pp 151–170.

Chapter 4
Simulating Residential Water Demand and Water Pricing Issues

P. Koundouri, M. Stithou, and P. Melissourgos

Abstract This chapter aims to simulate residential water demand in order to explore the importance of water for residential use. In addition, data on the water cost of supplying water in the residents of Asopos area from local distributors were collected. In order to capture the importance of water use specific parameters are examined and are used as indexes of water use. Some of these indexes are the population of the catchment, the number of households connected to the public water distribution system, m^3 of water consumption per year to cover household needs etc. The chapter closes with recommendations for designing and applying a program of measures for the efficient water resources management as described by Article 11 of Water Framework Directive (WFD).

Keywords Water for residential use • Tourism • Water pricing • Water service operators • Financial costs

1 Residential Area of Asopos RB

As it was emphasized in the previous chapters the first step in the implementation of the WFD is the economic characterization of water in the region requiring first of all that the economic significance of water is evaluated. This involves an assessment of the industrial, agricultural, residential and tourism water needs in the area. As the previous chapter covered the industrial and agricultural sectors the focus of this chapter is the residential and tourism. Hence, the current analysis will include information on the population connected to public water supply system *vs* those with self-supply, the total number of tourist days, employment and turnover in the tourism sector etc.

P. Koundouri (✉) • M. Stithou • P. Melissourgos
Athens University of Economics and Business, Athens, Greece
e-mail: pkoundouri@aueb.gr; stithoumavra@yahoo.com; pmelissourgos@aueb.gr

P. Koundouri and N.A. Papandreou (eds.), *Water Resources Management Sustaining Socio-Economic Welfare*, Global Issues in Water Policy 7,
DOI 10.1007/978-94-007-7636-4_4,

Table 4.1 Demographic data of Asopos RB

Prefecture	Municipality/Commune	Municipal departments-Residential areas	Population
Viotias	M. Plataion	Kaparelli, Lefktra, Plataiai, Loutoufi, Melissochori	4,114
	M. Thivaion	Thiva, Ambelochori, Neochoraki, Elaion	23,476
	M. Tanagras	Arma, Asopia, Kallithea, Tanagra, Panagia	4,076
	M. Dervenochorion	Dafni, Pyli, Panakto, Prasino, Skourta, Stefani	2,119
	M. Oinofiton	Oinofita, Dilesi, Ag. Thomas, Kleidi	7,869
	M. Schimatariou	Schimatari, Oinoi, Plaka, Dilesi	7,092
Attikis	M. Erithron	Erithres	3,105
	M. Oropou	Oropos, Skala, Markopoulo, Nea Palatia, Sikamino, Chalkoutsi, Nea Politeia	12,347
	M. Avlona	Avlona, Asprochori	5,085
	M. Malakasas	Malakasa, Milesi	1,292
Total	**Asopos River Basin**		**70,575**

Source: National Statistical Service of Greece (2001 Census) (http://www.statistics.gr/portal/page/portal/ESYE/PAGE-themes?p_param=A1602)

Regarding the residential area of Asopos RB, consisted of permanent residences, the selected indexes should reflect the economic importance of domestic water use in the specified area. Therefore, it is useful for these indexes to offer full data not only for the demand but also for the supply of water. The employed indexes for the evaluation of the economic importance of water supply in the residential case study area are the following:

i. *Permanent population*

In order to calculate the permanent residents of Asopos RB, the Municipalities/Communes that fall geographically within the RB were considered. The geographical distribution of Asopos RB was defined according to the National Programme for the Management and Protection of Water Resources (MoEPPW – NTUA 2008). For the Municipalities/Communes that don't fall wholly within the territory of Asopos RB only the permanent population of the subsequent Municipal Departments and urban localities that are included in Asopos RB was calculated. Considering these the total permanent population of the RB is 70,575 residents (Table 4.1).

ii. *Permanent main dwellings*

According to the National Statistical Service of Greece the term "Regular dwelling" is defined as the permanent and single-family housing which is consisted of at least one room and is used as the residence of one household. Occupied main regular dwelling is the residence in which the household resides during most of the year and it was occupied the day that census took place (17th–18th March 2001). Non regular dwellings are constructions made by crude materials of little value without a predefined plan (huts, trailers/caravans, storehouses etc.), which are not regular dwellings but were occupied the day of census. A regular dwelling within a collective residence is the regular dwelling which is within a residence that is to be occupied or to service many individuals or groups of people. Examples are hotels, hospitals, jails, elderly homes or the residence of the director of the collective residence or of the personnel etc. Vacation residence is the regular dwelling which is normally further away from the main dwelling of the household and is used for resting or recreation during summer or on a seasonal basis. Second residence is the residence that is used by the household in parallel with the main dwelling without been a vacation residence.

Regarding the above terminology it is concluded that within Asopor RB there are 23,000 households (Table 4.2) which are permanently connected with the public system of water supply. The rest of residences (second or vacation residences, residences for rent or sale, non regular dwellings or regular which is part of a collective residence) are not counted in as it is likely that they are not connected with the public system of water supply or they are not permanent residences. It should be noted that the total number of residences (43,223) is the total of regular, non regular and regular within collective residence residences of the Municipal/Commune departments and localities (town or village) of these Municipalities/Communes that belong geographically to the Asopos RB.

iii. *Employment*

According to the Labour Force Survey of the National Statistical Service of Greece, the population of working age is divided into three mutually exclusive and exhaustive groups – persons in employment, unemployed persons and inactive persons. The employed are persons aged 10 years or older,[1] who during the reference week had worked, even for just one hour, for pay or profit or they have worked in the family business, or they were not at work but had a job or business from which they were temporarily absent. Unemployed are persons aged 10 years and over[2] who were without work during the reference week (they were not classified as employed), were currently available for work and were either actively seeking work in the past 4 weeks or had already found a job to start within the next 3 months. They are distinguished between persons who had lost their job (for any reason) and young people who are looking for a job for the first time. The employed and unemployed people form the economically active population. Inactive are those persons who neither classified as employed nor as unemployed.

Regarding the above terminology and data of Table 4.3, the total economically active population (labor force) of Asopos RB in 2001 was 31,764. Of these persons 28,837 are employed (90.8 % of the total economically active population) and 2,927 are unemployed (9.2 % of the total economically active population). It is worth noting that for the rest of the country in September 2009 the respective unemployment rate was 9.1 %, while in particular for the Prefecture of Sterea Ellada it was 9.7 %. Furthermore, from the total of employed people the 19.8 % are employed in thevprimary sector, the 28 %

[1] This range is kept for comparative reasons with the previous census.

[2] Op.cit. 4.

Table 4.2 Dwellings in Asopos RB

Prefecture	Municipality/Commune Urban (Ur) Agricultural (Ag) Lowland(Lo) Semi-mountainous (S-m)	Total dwellings	Regular dwellings Private-Public sector ownership and LEPrL[a]- LEPuL[b] Main occupied residence	Second or vacation residence	Residence for sale, rent etc.	Non regular dwellings	Regular dwellings within collective residence
Viotias	Plataion (Ag, S-m, Lo)	2,599	1,541	927	97	34	0
	Thivaion (Ur, ΑΓ, Π)	9,096	7,382	436	1,221	47	10
	Tanagras (Ag, Lo)	1,344	1,035	167	79	61	2
	Dervenochorion (Ag, S-m)	1,134	705	368	55	4	2
	Oinofiton (Ur, Ag, Lo)	4,621	2,664	1,597	333	26	1
	Schimatariou (Ur, Lo)	4,974	2,259	2,429	222	54	10
Attikis	Erithron (Ur, Lo)	1,561	1,121	328	112	0	0
	Oropou (Ag, Ur, Lo, S-m)	15,492	4,630	9,487	1,339	17	19
	Avlona (Ur, S-m)	1,585	1,213	225	93	10	44
	Malakasas (Ag, Lo)	817	450	303	56	5	3
Total (%)	**Asopos RB**	**43,223 (100 %)**	**23,000 (53.2 %)**	**16,267 (37.6 %)**	**3,607 (8.4 %)**	**258 (0.6 %)**	**91 (0.2 %)**

Source: National Statistical Service of Greece (2001 Census) (http://www.statistics.gr/portal/page/portal/ESYE/PAGE-themes?p_param=A1602)

[a]Legal Entity of Private Law (Company, Union etc.) (LEPrL)

[b]Legal Entity of Public Law (Municipality, Social Insurance Organization etc.) (LEPuL)

Table 4.3 Economically active and inactive population in Asopos RB

		Economically active population							
			Employed					Unemployed	
Prefecture	Municipality/Commune	Total	Total	Primary sector	Secondary sector	Tertiary sector	Did not state economic activity	Total	Economically inactive
Viotias	M. Plataion	1,829	1,723	883	384	441	15	106	2,045
	M. Thivaion	10,613	9,347	1,610	2,680	4,081	976	1,266	10,279
	M. Tanagras	2,193	2,077	1,157	380	447	93	116	1,595
	M. Dervenochorion	870	790	269	237	265	19	80	1,033
	M. Oinofiton	3,951	3,649	448	1,525	1,102	574	302	3,241
	M. Schimatariou	3,428	3,156	548	1,161	985	462	272	3,006
Attikis	M. Erithron	1,178	1,073	194	243	586	50	105	1,650
	M. Oropou	5,070	4,667	770	1,314	2,358	225	403	6,218
	M. Avlona	2,068	1,833	379	800	614	40	235	2,649
	M. Malakasas	564	522	31	167	313	11	42	750
Total (%)	**Asopos RB**	**31,764 (100 %)**	**28,837 (90.8 %)**	**6,289 (19.8 %)**	**8,891 (28 %)**	**11,192 (35.2 %)**	**2,465 (7.8 %)**	**2,927 (9.2 %)**	**32,466**

Source: National Statistical Service of Greece (2001 Census) (http://www.statistics.gr/portal/page/portal/ESYE/PAGE-themes?p_param=A0101)

in the secondary sector, and the 35.2 % in the tertiary sector while the 7.85 of employed people did not report a sector of economic activity.

The total economically active and inactive population of Asopos RB (31,764 + 32,466 = 64,230) is calculated, as in the previous section, considering the permanent population of the individual Municipal/Commune departments and localities of Municipalities/Communes which are geographically included in Asopos RB.

iv. *Competent bodies and method of water supply and sewerage*

Competent bodies responsible for water supply and sewerage were also presented in Chap. 3 and therefore only a short reference is made here. It is reminded that in Greece the provision of water supply and sewerage is treated as a public service and that there are 214 companies of water supply and sewerage services in the country. The competent bodies/companies of water supply and sewerage are divided into three categories (Safarikas et al. 2006; Tsagarakis et al. 2003):

(i) In the cities of Athens and Thessaloniki there are private, public and nonprofit companies of water supply and sewerage which posses a 20 year concession. These are the Athens Water Supply and Sewerage Company (EYDAP S.A) and the Thessaloniki Water Supply & Sewerage Co (EYATH S.A). They come under the jurisdiction of the MoEECC which approves their water pricing policy. Overall, the objective of the above companies is the supply of water and sewerage services in their territory as well as the research, construction, establishment, operation, exploitation/use, management, maintenance, expansion and renewal of the systems of supply and sewerage.

(ii) In the cities with more than 10,000 residents the Municipal Enterprises for Water Supply and Sewerage (DEYA) operate as private companies that belong to the Local Government according to the Law 1069/80. However, there are towns with a population of less than 10,000 residents in which DEYA are located. These companies are LEPrL, of public interest and of specific mission and are in accordance with the rules that govern the private economy. Every DEYA, has as objective the supply of water and sewerage to the consumers while it is responsible for: the quality of water, the early respond to a likely water shortage, the good condition of the water supply and sewerage system, the construction of water supply projects etc.

(iii) In the rest of the areas (Towns/Municipalities with less than 10,000 residents) the competent bodies of water supply and sewerage services are the Municipalities. These Municipalities are about 830 and it is estimated that they serve about 12 % of the country's population.

Focusing in the area of Asopos RB the above three mentioned separate systems of water supply and sewerage and combinations of these are found. More specifically, for the individual Communes and Municipalities of the Asopos RB the following competent bodies of water supply (Table 4.4), with their respective points of abstraction and methods of water treatment and supply, are observed:

v. *Water pricing policy for water supply for residential use*

As it has been also emphasized pricing of water consumption is very important since it is related to the credibility and quality of water supply as well as to the

Table 4.4 Competent bodies, points of abstraction and methods of water supply

Prefecture	Municipality/ Commune	Competent body of water supply	Abstraction points	Method of water treatment and supply
Viotias	M. Plataion	M. Plataion	Processed water by EYDAP S.A	Chlorination
	M. Thivaion	DEYA Thivas	Morno's aqueduct and 16[a] boreholes of EYDAP or EYATH	Refinement → Chlorination
	M. Tanagras	M. Tanagras	4 municipal boreholes (P. Hilias, Kallithea, Gipedo, P. Hilias)	Chlorination
	M. Dervenochorion	M. Dervenochorion	Processed water by EYDAP S.A and Dafni's borehole	Refinement → Chlorination
	M. Oinofiton	M. Oinofiton	Raw water from Morno's Reservoir through EYDAP and local Municipal boreholes	Refinement
	M. Schimatariou	M. Schimatariou	Raw water from Iliki lake through EYDAP which is που additionally mixed with the water of two municipal boreholes (K.Ygeias - Trita Dilesiou)	Pre-chlorination → Flocculation/ Coagulation → Refinement → Post-chlorination → Storage
Attikis	M. Erithron	M. Erithron	Processed water by EYDAP S.A	Disinfection → Flocculation/ Coagulation → Sedimentation tank → Filtering → Post-chlorination
	M. Oropou	M. Oropou	Mavrosouvala's Association of Water Supply and Drilling Exploitation (15 boreholes, 100,000 m^3/day)[b]	Chlorination
	M. Avlona	M. Avlona	4 boreholes with 3 pumping stations: Skalezas, Pylas, Asprochoriou with plumbing supply of 100 m^3/h, 80 m^3/h, 60 m^3/h respectively	Refinement → Chlorination
	M. Malakasas	M. Malakasas	Municipal boreholes of Mavrosouvala's Association of Water Supply and Drilling Exploitation	Chlorination

[a]They were operation till 1st of September 2009 where they stopped to supply the water network of the Municipality due to the problem of the findings of hexavalent chromium

[b]LEPuL which was established having as objective the supply of water in the neighboring Communes and Municipalities with a charge of 0.12 €/m^3

possibility of the development of new water management systems in order to satisfy future, quantitative and qualitative needs.

As reported in previous chapter on industrial water use, water suppliers such as EYDAP S.A, EYATH S.A, DEYA and Municipalities, follow a different water pricing policy which is ought to the fact that they differ in terms of operation and financial data. EYDAP S.A and EYATH S.A, as privatized state owned and nongovernmental companies, set the prices which should be approved by the state and should cover their cost of operation. The pricing that EYDAP S.A employs defines ten categories of water use and users and imposes successive/escalating or standard charge and a monthly fixed charge depending on 3 months consumption.

For the domestic water use that includes residential, urban areas a graduated charge is imposed such as: for monthly consumption 0–5 $m^3 \rightarrow 0.4138$ €/m^3, for 5–20 $m^3 \rightarrow 0.6471$ €/m^3, for 20–27 $m^3 \rightarrow 1.8566$ €/m^3, for 27–35 $m^3 \rightarrow 2.5992$ €/m^3, for more than 35 $m^3 \rightarrow 3.2357$ €/m^3. In addition, the water bill of EYDAP S.A. discerns among four categories of users for the right of using the sewers (common consumers, public-municipal, industrial, charity) who are charged as a percentage of the water price (e.g., common consumers: 75 % × value of consumed water). Finally, the bill includes charges that take the form of the Value Added Tax (V.A.T.) of 9 % of the value of the consumed water and 19 % of the rest of charges as well as 1 % of the value of consumed water as a contribution to the fund for personnel insurance.

Every DEYA which operates as a private company sets the prices which have to be approved from the City Council and to cover the operational cost. Hence, for the calculation of the operational cost and the implementation of water pricing factors such as salaries, subsidies, taxes, depreciation of capital etc. are considered. As such

In particular, Thiva's DEYA (the only DEYA in Asopos RB) applies a water policy that differentiates across the Municipal Departments that belong to Thiva's Municipality. Therefore, regarding the Municipal Department of Thiva for year 2008, Thiva's DEYA discerns six categories of water users and uses and imposes an escalated or fixed charge according to the type and amount of the consumption and an annual fixed cost of water supply (€60). The water bills are every 2 months as follows: for 2 months consumption from 0 to 30 $m^3 \rightarrow 0.45$ €/m^3, for 31–60 $m^3 \rightarrow 0.60$ €/m^3, for 61 m^3 and over$\rightarrow 1.00$ €/m^3. In addition, DEYA's water bill for the Municipal Department of Thiva discerns two categories of users regarding the right for sewer and use of sewer (domestic use, industrial use). In particular, regarding domestic users the percentage charge is the following: fee for sewer use→70 % of the value of the consumed water, drainage fee →90 % of the value of the consumed water.

For the rest of the Municipal Departments of the Municipality of Thiva that are included in Asopos RB (Eleonas, Neochoraki, Ambelochori) the DEYATh discerns into two categories of use (domestic and professional use) of a common however pricing and applies a 4 month billing. In particular, the water bills and the graduated charge is the following:

Municipal Department of Eleonas: for consumption from 0 to 50$m^3 \rightarrow 0.15$€/m^3, for 51–90 $m^3 \rightarrow 0.18$ €/m^3, for 91 m^3 and over$\rightarrow 0.30$ €/m^3. Annual fixed fee of water supply→21.00 €

Municipal Department of Neochoraki: for consumption from 0 to 50 m^3→0.18 €/m^3, for 51–90 m^3→0.24 €/m^3, for 91 m^3 and over→0.30 €/m^3. Annual fixed fee of water supply →21.00 €

Municipal Department of Ambelochori: for consumption from 0 to 50 m^3→0.22 €/m^3, for 51–90 m^3→0.32 €/m^3, for 91 m^3 and over 0.37 €/m^3. Annual fixed fee of water supply →30.00 €

Finally, in all the Municipal Department of the Municipality of Thiva (Thiva, Eleona, Neochorakiou, Ambelochoriou) a charge of special fee 80 % is imposed to all the bills excluding the State consumption and that of LEPuL and of the Municipality of Thiva. The value of the consumed water bears a V.A.T. 9 % while the rest of the charges (fixed fees of water supply-sewerage-use of sewer, special fee 80 %, materials, tasks etc.) bear a V.A.T. 19 %.

For the Municipalities/Communes that operate as the competent bodies of water supply and sewerage services, the water pricing policy that is applied is very heterogeneous. Every Municipalities/Commune applies a different pricing in water consumption by employing different criteria and by setting different financial, qualitative and quantitative targets. As a result, across the Municipalities/Communes of Asopos RB which operate as competent bodies of water supply for the domestic consumers, it is observed a different escalation regarding the consumed cubic meters of water, different charge for similar scales and different fixed charge (Table 4.5). More analytically the water pricing policy that has been followed in the Municipalities/Communes for the pricing of domestic water consumption is as follows:

M. Schimatariou
0–60 m^3→0.60 €/m^3
61–120 m^3→0.90 €/m^3
121–180 m^3→1.46 €/m^3
181–240 m^3→1.85 €/m^3
241 m^3 and over→2.51 €/m^3
Fixed fee→15.68 €/6 months

M. Erithron
0–20 m^3→0.50 €/m^3
21–50 m^3→0.60 €/m^3
51–80 m^3→0.70 €/m^3
81 m^3 and over→0.80 €/m^3
Fixed fee→10 €/4 months

M. Avlona
0–40 m^3→0.25 €/m^3
41–150 m^3→0.35 €/m^3
151–300 m^3→0.45 €/m^3
301–500 m^3→0.80 €/m^3
Fixed fee →12 €/6 months

M. Plataion
0–100 m^3→0.70 €/m^3
101–180 m^3→0.75 €/m^3
181 m^3 and over→1.00 €/m^3
Fixed fee →15 €/6 months

M. Oinofiton
0–120 m^3→0.29 €/m^3
121–999999 m^3→0.44 €/m^3
Fixed fee →8.80 €/6 months

M. Oropion
0–120 m^3→ 30 € fixed fee
121–200 m^3→0.70 €/m^3
201 m^3 and over →1.00 €/m^3
Fixed fee→30 €/6 months

M. Tanagras
0–300 m^3→0.36 €/m^3
300 m^3 and over →0.54 €/m^3
Fixed fee→13.21 €/6 months

M. Dervenochorion
0–50 m^3→ 1.00 €/m^3
51 m^3 and over→1.20 €/m^3
Fixed fee→8.00 €/6 months

M. Malakasas
The required data of water pricing were not provided

Table 4.5 Water pricing in Asopos RB

Prefecture	Municipality/ Commune	Competent body of water supply	Fixed fee of water supply	Charge of the 100nd m³ (€/m³)
Viotias	M. Plataion	M. Plataion	15 €/6 months	0.70 €/m³
	M. Thivaion	DEYA Thivas	60€/years[a], 21€/years[b], 30€/years[c]	1.00€/m³[a], 0.30€/m³[b], 0.37€/m³[c]
	M. Tanagras	M. Tanagras	13.21 €/6 months	0.36 €/m³
	M. Dervenochorion	M. Dervenochorion	8.00 €/6 months	1.20 €/m³
	M. Oinofiton	M. Oinofiton	8.80 €/6 months	0.29 €/m³
	M. Schimatariou	M. Schimatariou	15.68 €/6 months	0.90 €/m³
Attikis	M. Erithron	M. Erithron	10 €/4 months	0.80 €/m³
	M. Oropou	M. Oropou	30 €/6 months	0–120 m³→30 €
	M. Avlona	M. Avlona	12 €/6 months	0.35 €/m³
	M. Malakasas	M. Malakasas	[d]	[d]
Average	**Asopos RB**	–	**30.26 €/year**	**0.60 €/m³**

[a]For the Municipal department of Thiva
[b]For the Municipal departments of Eleona and Neochorakiou
[c]For the Municipal department of Ambelochoriou
[d]The Municipality of Malakasa did not provide us with the required data

vi. *Water consumption*

The 5–20 % of water consumed on a daily bases in all the main water uses (industrial, domestic, irrigation) is used from the households through domestic water use. The domestic use of water is classified into indoor and outdoor water use. While the indoor water use is mainly dependent on the habits of the residents, the outdoor water use depends mainly on the size of the territory, its climate etc. From this percentage of domestic water use the 40 % of water is discharge through the toilet. The rest of the quantity is used as follows: 25 % is used in the bathroom, 20 % for cloth and dish washing, 10 % in the kitchen and 5 % for cleaning. In Greece, 90 % of the households have access to a network of water supply compared to 30 % during the 1950s. The use of water for water supply has increased by 45 % compared to 1980 and is still increasing. This increase has been related to the building, the use of appliances such as washing machines and contemporary facilities such as gardens, swimming pools etc.[3]

The domestic water consumption reflects successfully the size of water demand. In particular, for the Communes and Municipalities of Asopos RB the following annual water consumptions, reported in Table 4.6, in 2008 are observed.

[3]http://www.watersave.gr/site/index.php

Table 4.6 Annual water consumption in Asopos RB (2008)

Prefecture	Municipality/Commune	Annual water consumption (m^3)
Viotias	M. Plataion	305,786
	M. Thivaion	2,100,000
	M. Tanagras	383,501
	M. Dervenochorion	220,000
	M. Oinofiton	775,794
	M. Schimatariou	1,650,924
Attikis	M. Erithron	321,925
	M. Oropou	4,000,000[a]
	M. Avlona	567,000
	M. Malakasas	170,000[b]
Total	**Asopos RB**	**10,494,930**

[a]Approximate estimation, since the Municipality of Oropos does not have data on water consumption and the Association of Water and Drilling Exploitation of Mavrosouvala provided estimations of water consumption for 2008 since the system of tele-metering of the pumping station was not yet installed

[b]Estimation according to the population and the consumption of other similar Municipalities, since the Municipality of Malakasa did not provide the required data

2 Area of Tourism Activity

Regarding the tourism area of Asopos RB, including vacation/holiday homes and tourist infrastructure or hotels, indexes and parameters are selected that show clearly the economic importance of water supply in the tourism sector. However, due the special characteristics of the area, the limited touristic development but also the lack of necessary data, the indexes and parameters that potentially can be used are limited. Nevertheless, the conclusions that are reached for the economic importance of water supply in the residential area of Asopos, could be hypothetically be valid and expand to the touristic area of Asopos RB.

This is ought to the fact that tourism in the area is mostly characterized by holiday-makers that reside in holiday homes rather than tourists that use for their accommodation tourist units and hotel resorts. This is also due to the proximity of the area to urban places such as Athens which is a motivation for the construction of vacation and second residences. On the other hand, the industrial development of the area in combination with the environmental degradation that it brings is a disincentive for the development and location of touristic units and hotels.

Concluding, the tourist area of Asopos RB is mainly consisted of vacation or second residences. For that reason, the indexes of estimating the economic importance of water supply in the residential area such as water suppliers and water pricing policy can be also employed in the touristic study area. In addition, for a clearer characterization of the economic significance of water supply in the touristic area the following parameters are employed:

Table 4.7 Holiday/vacation – Second residences of the municipalities of Oropos and Tanagra

Municipality of Oropos	Holiday/vacation or Second residences	Municipality of Tanagra	Holiday/vacation or Second residences
M.D. Markopoulou	2,158	M.D. Armatos	13
M.D. Neon Palation	2,314	M.D. Asopias	39
M.D. Skalas	3,649	M.D. Kallitheas	23
M.D. Sykaminou	857	M.D. Tanagras	92
M.D. Oropou	509	–	–
Total	**9,487**	**Total**	**167**

i. *Holiday/vacation – Second residences*

According to Table 4.2 where information of dwellings in Asopos RB is reported, the number of vacation or second residences is 16,267 and is the 37.6 % of the total residences in Asopos RB. It is worth noting that from the total of second or vacation residences most of them are located in the coastal areas of the Municipalities. Table 4.7 reports the data of the Municipality of Oropos and the Municipality of Tanagra.

Hence, from the total of 16,267 vacation or second residences in Asopos RB, the 9,487 are located in the Municipality of Oropos and in particular 8,121 are located in the coastal areas of the Municipal Department (M.D.) (Markopoulo, Nea Palatia, Skala). On the contrary, only 167 vacation or second residences are located in the Municipality of Tanagra. This fact shows the high positive correlation that exists in the area between holiday residences and coastal areas (without industrial development) and at the same time, the high negative correlation that is observed in the area between vacation residences and lowland industrial areas.

ii. *Water consumption during May-September*

The demand for water is not equally distributed during the year. Especially during the summer season (May-September) the demand and consumption of water increases considerably due to the specific climate conditions that prevail during that season (e.g., heat, drought). In specific areas of Asopos RB, water consumption during summer months increases doubles compared to consumption during the rest of the months. In addition, the positive correlation between holiday homes and coastal areas demonstrates the high positive correlation between water consumption and summer in such areas. The existence of many holiday homes/cottages, coastal or not, has as a result for the population of the area to double or triple during summer and hence to observe high water consumption. In order to explore the contribution of the touristic study area to the water consumption it is useful to record the total water consumption during the summer months. Specifically, water consumption for the basin of Asopos during May-September 2008 is presented in Table 4.8.

Table 4.8 Water consumption during May–September

Prefecture	Municipality/Commune	Water consumption (m^3)
Viotia	M. Plataion	200,258
	M. Thivaion	1,080,000
	M. Tanagras	200,000[a]
	M. Dervenochorion	120,000
	M. Oinofiton	243,526
	M. Schimatariou	832,557
Attiki	M. Erithron	173,608
	M. Oropou	2,300,000[a]
	M. Avlona	226,600
	M. Malakasas	90,000[a]
Total	**Asopos RB**	**5,466,549**

[a]These specific measurements are estimates due to lack or no access to monthly data of consumption. Estimates are calculated according to the population and consumption of similar districts

3 Conclusions and Policy Recommendations

In this chapter an attempt was made to evaluate the economic significance of water supply in the residential and touristic area of Asopos RB. The economic evaluation of the significance of water use is the first step of the first stage of WFD implementation. In this context and regarding our particular case study data provided by the responsible water providers (Municipalities, DEYA) were used to create informative indexes. It should be also noted that other important parameters such as "income" could not be employed due to lack of necessary data or of the possibility to be published by the competent bodies.

Overall, regarding the residential area the general conclusions are that:

- The total permanent population of Asopos RB is 70,575 residents which shows the importance for qualitative and optimal water supply in the area. The total permanent population could be also taken into consideration as long as other parameters and data that concern the area such as urban wastewater, urban water consumption, sewerage etc.
- The total number of dwellings in Asopos RB reaches 43,223 residences from which at least 43,000 residences are considered to be connected to or going to be connected to the water supply network. In addition, the 23,000 regular residences are permanent residences and therefore it is regarded that are permanently connected with the network of water supply and consume water during the whole year. Finally, it is concluded that the main-permanent residences (23,000) combined with the vacation-second regular residences (16,267) of Asopos RB put the higher pressure on the total domestic water use.
- The economically active population of Asopos RB reaches the 31,764 (employed and unemployed) of which the 35.2 % is occupied in the tertiary sector, the 28 % in the secondary sector, the 19.8 % in the primary sector and the 9.2 % are

unemployed. The economically non-active population is 32,466. However, due to the intense industrial development of some areas of the Asopos RB like that of Oinofita, Schimatari and Tanagra, in the economically active population of Asopos RB it should be also included the people who are working in the area but reside in other areas. Exemplary is the Municipality of Oinofita where on a daily basis 15,000 industrial workers are transferred excursing therefore a pressure on water consumption.

- The competent bodies of water supply in Asopos RB include one DEYA and nine Municipalities/Communes. However, some of these competent bodies provide their services in collaboration with EYDAP S.A. The way of water supply, treatment and the abstraction points differ among the competent bodies and that justifies different quality, available quantity and charge of water.
- Due to the different way of operation of the competent water supply bodies and the different financial data that are to be included in the estimation of the charge, what is observed is a diversified water pricing policy within Asopos RB. This diversification in water pricing is not only observed among the different systems of water supply (e.g., DEYA and Municipalities/Communes) but it is also observed within the same system of water provision (e.g., Municipalities/Communes). Since WFD asks for water pricing that also considers the environmental and the resource cost of water use, the competent bodies are called to adopt a water policy which is not solely based on the financial cost and which promotes cost recovery. It seems that the change in the water bills is the only and necessary solution of the problem of the partial recovery of the real cost of water supply. In this context, the charges need to reflect the real cost and should not only include the abstraction cost, the cost of transfer and network maintenance but also the environmental and resource cost. This does not necessarily mean higher charges for all sectors. The water pricing policy of the competent bodies should be primary socially equal. This can be achieved through the graduated charges, reliefs for the most vulnerable social groups and through the application of the principal "the polluter pays" so as the real cost of water not to burden unfairly and wholly the society. Furthermore, due to the special characteristics of Asopos RB (serious industrial pollution) the adjustment of the water pricing should be applied on a common and uniform base. As a result, in order to achieve a full cost recovery of the real cost of water in the whole area of Asopos, the competent bodies should apply a uniform water policy that includes common graduated charges and fixed fees. In addition, the common water pricing in Asopos RB should also include charges to the industries in order to gather the necessary sources for the construction of remediation works of Asopos' ecosystem and rehabilitation of the area. In order to implement the above a necessary condition is the revision and amendment of the legal framework. The creation of the new legal framework which will be applied in the area should include, apart from the adjustments in the water pricing, the necessary measures of rehabilitation and protection of the area as well as the legal and monetary penalties in case of pollutant related activity. The current water pricing that is applied in the area

of Asopos presents an average fixed fee of water supply of €30.26 per year and average charge of the 100th m^3 of €0.60 per m^3.

- The water consumption in the residential area of Asopos reflects the need and the demand for domestic water use. However, the recorded measurements cannot be considered absolutely accurate. In reality the specific measurements should be considered of less magnitude in m^3 than the real water consumptions. This happens because the reported values are either measurements based on water metering installed in the households or measurements based on water metering installed in water abstraction points. These measurements are normally smaller that the real water consumptions since they cannot include any water leaks as a result of damages in the network, of leaks during transfer, abstraction etc. The annual consumption in Asopos RB for 2008 is 10,494,930 m^3. However, it should be noted that this number is not absolutely accurate since water simulations in few cases took place due to lack of data.

Regarding the touristic area of Asopos the most important conclusions are the following:

- The tourism development including hotels, accommodation to rent, camping and other tourist units, in the area is limited and takes place only in specific districts. This is mainly caused by the intense industrial development and the pollution that it brings creating disincentives for the touristic development. On the contrary, the tourist development in the area in terms of holiday and vacation residences is of particular interest. There are 16,267 holiday or second residences which correspond to the 37.6 % of the total of residences in Asopos. However, these residences are not equally distributed across the district but are mainly concentrated in the coastal areas. Obviously the existence of a high number of vacation or second residences puts more pressure on the demand for water consumption.
- Water consumption during summer and particularly during May-September, shows a rising trend. This fact is partly due to the special climatic conditions of the specific time period and to the increase of water consumers during that specific time. For example, the Municipality of Schimatari that reports a consumption of 89716.2 m^3 in February 2008, in September 2008 it equals 213,090 m^3. The total water consumption in Asopos RB during May-September 2008 is 5,466,549 m^3 and is about 50 % of the annual consumption. A response to moderate water use during summer in the area could be the implementation of a seasonal graduated water charge by the competent bodies of water supply. According to the seasonal diversified percentages or seasonal percentages, the charges are higher during peak seasons (summer) and lower during periods of low demand (winter).

Concluding, this chapter with a focus on the residential water users presented an assessment of the financial cost of water service operators. Financial costs are likely to need a number of adjustments before economic costs can be identified: typically, adjustments for transfers and taxes but also adjustment for the environmental and resource costs caused by water service operators' activities. Environmental cost attributed to the water service operators in the area is not available. However, there are costs which water service operators incur because they operate within a polluted environment or an environment from which raw water resources are currently depleted by other water users (EUREAU 2004). Hence, the water service operator may incur additional water treatment costs or additional cost of transportation.

In any case, the costs of operating within a polluted environment should not be recovered from water service users but from the polluters who caused them and in case of Asopos is the industrial and then the agricultural sector. Hence, as mentioned previously a program of measures should be considered which would charge industrial sector, as in the case of any other user, respectively with the total cost that generates. As it will be presented in Chaps. 6 and 7 the total cost of for example ecosystem degradation is estimated from €882,200 to €2,690,000 per year using different methods of estimation. Finally, it is expected that full cost recovery through charges and measures will motivate the adoption of green investments and the decrease of pollution to the social optimal level.

References

European Union of National Associations of Water Suppliers and Waste Water Services (EUREAU). (2004). *Water Framework Directive: Determination of cost recovery.*

Hellenic Republic. Ministry of Environment Physical Planning and Public Works (MoEPPW)-National Technical University of Athens (NTUA), Department of Water Resources and Environmental Engineering. (2008). *National programme for the management and protection of water resources.*

Safarikas, N., Paranychianakis, N. V., Kotselidou, O., & Angelakis, A. N. (2006). Drinking water policy in the frame of the Directive 2000/60/EC with emphasis on drinking water prices. *Water Science and Technology: Water Supply, 5*(6), 243–250.

Tsagarakis, K. P., Paranychianakis, N. V., & Angelakis, A. N. (2003). Water supply and wastewater services in Greece. In S. Mohajeri, et al. (Eds.), *European water management between regulation and competition, aqualibrium project, EU-Directorate-general for research, global change and ecosystems*, B-1049, Brussels, Belgium, pp. 151–170.

Chapter 5
An Econometric Analysis of Agricultural Production, Focusing on the Shadow Price of Groundwater: Policies Towards Socio-Economic Sustainability

P. Koundouri, O.G. Dávila, Y. Anastasiou, A. Antypas, T. Mavrogiorgis, A. Mousoulides, M. Mousoulidou, and K. Vasiliou

Abstract The focus of this chapter is on the agricultural sector in the Asopos catchment as it has a significant impact on the status of water in the area. In particular, the aim of the chapter is to estimate the farmers' valuation of groundwater's shadow price for the region of Asopos. In order to achieve that, an agricultural micro-economic data-set from the catchment has been collected through the use of a detailed agricultural questionnaire. As it will be explained in the chapter, the questionnaire focuses on collecting information regarding cultivations, production structures and use of groundwater for irrigation. The objective of the micro-econometric analysis is to uncover patterns of groundwater use and farm efficiency. The chapter presents the derived estimates that make possible the analysis of the impact of different economic policies, – which will be used for the implementation of an optimal, sustainable and integrated water policy – on farmers' profits and social welfare. The chapter finishes with policy recommendations based on the principle of socio-economic sustainability that assures both economic efficiency of farms and concludes with the estimation of groundwater for irrigation shadow price and how this can be used in the design of pumping taxes to reduce pollution and to increase farms efficiency.

Keywords Shadow price of groundwater • Groundwater use • Farm efficiency • Irrigation • Agricultural production

---

P. Koundouri (✉) • O.G. Dávila • Y. Anastasiou • A. Antypas • T. Mavrogiorgis
M. Mousoulidou • K. Vasiliou
Athens University of Economics and Business, Athens, Greece
e-mail: pkoundouri@aueb.gr; osiel.davila@soas.ac.uk; yanastasiou@aueb.gr; aantypas@unipi.gr; tmavrogiorgis@aueb.gr; mmousoulidou@aueb.gr; kvasiliou@aueb.gr

A. Mousoulides
Athens University of Economics and Business, Athens, Greece

Intercollege Larnaca, Larnaca, Cyprus
e-mail: mousoulides.a@adm.lar.intercol.edu

P. Koundouri and N.A. Papandreou (eds.), *Water Resources Management Sustaining Socio-Economic Welfare*, Global Issues in Water Policy 7,
DOI 10.1007/978-94-007-7636-4_5,

1 Introduction

The project aims to evaluate impacts of implementing the EU water directive requiring full cost recovery by 2015. The use of groundwater by farmers located above an aquifer is a very common situation in irrigated agriculture. In such case an estimate of groundwater's shadow price can be used to value the stock of water in green accounting calculations, to help design pumping taxes in order to mitigate externalities associated with groundwater use (Howe 2002). The Asopos watershed has multiple uses and has significant industrial water pollution and aquifer depletion. Farmers are using pumps illegally. There are concerns for contamination of agricultural products from polluted water (potatoes for example). Based on the data provided in Appendix, it appears that wheat, potato and olive production are distinct, monoculture systems (with potato irrigated). These crops represent about 80 % of total cultivated area. The other major land use is a mix of crops that are mostly irrigated vegetables (about 20 % of the cultivated area). Thus, the overall population could be modeled as four strata: (1) rainfed wheat farms; (2) rainfed olive farms; (3) irrigated potato farms; (4) irrigated vegetable/mixed crop farms. More information on the local farms features is presented in Appendix. The following sections will deal with the estimation of groundwater's shadow price. In Sect. 1, the distance function in production theory is analyzed. Section 2 presents the econometric specification, empirical estimation, and results of the estimation of groundwater's shadow price.

2 The Distance Functions in Production Theory

The seminal work by Shephard (1970) provided the theoretical foundations on distance functions in production theory.[1] Grosskopf and Hayes (1993), Färe et al. (1993) and Coelli and Perelman (2000) conducted empirical applications in order to compute shadow prices of either inputs or outputs in various regulated industries. Among the advantages of using distance functions we can distinguish the following:

(a) Data on prices is not required in order to conduct the parameters estimation, only quantity data is necessary.
(b) No behavioral hypothesis (like profit maximization or cost minimization) are imposed on these models and
(c) Firm-specific inefficiencies can be calculated using distance functions.

These are great advantages when tackling water management problems since reliable price data for natural resource inputs is scarce. When firms are heavily regulated they often have a diversity of objectives, and due to regulation or non-optimal management of natural resource industries inefficiencies may arise (in this case, inefficiencies in the use of groundwater). Following Färe and Grosskopf (1990), a dual

[1] It should be noted that Färe et al. (1994) and Färe and Primont (1995) Färe et al. conducted extensions to Shephards' work.

Shephard's lemma is employed to retrieve firm and input specific shadow prices by using a Shephard's input distance function to characterize technology rather than a cost function. It is considered that an input approach is appropriate in the analysis of the agriculture sector because the managers are likely to have more discretionary control over inputs rather than outputs (Koundouri and Xepapadeas 2004). The restricted input distance function for the *i*th agricultural firm (D_i^R) is defined as:

$$D_i^R\left(Y_i, \boldsymbol{X}_i^P, \boldsymbol{X}_i^W; W_i, H, T\right) = \max\left\{\phi_i > 0 : \frac{\boldsymbol{X}_i}{\phi_i} \in L_i\left(Y_i; W_i, H, T\right)\right\} \quad (5.1)$$

Where X_i denotes firm-specific vector of *m* input quantities ($X_i = (X_i^P, X_i^W) \in R_+^m$), $L_i(Y_i; W_i, H, T) = \{X_i \in R_+^m : X_i \text{ can produce } Y_i \in R_+\}$ denotes the set of all input vectors which can produce the output vector ($Y_i \in R_+$); and ϕ_i measures the proportional reduction in all ($X_i \in R_+^m$) that brings the *i*th firm to the frontier isoquant. Given the restricted cost function in Eq. 5.2, Shephard (1970) showed that the restricted input distance function may also be obtained as a price minimal cost function as shown in Eq. 5.3.

$$\begin{aligned} C_i^R &= \left(Y_i, \boldsymbol{P}_p, \boldsymbol{P}_w; W_i, H, T\right) \\ &= \min_{X_i^P, X_i^W}\left\{\boldsymbol{P}_p \boldsymbol{X}_i^P + \boldsymbol{P}_w \boldsymbol{X}_i^W : D_i^R\left(Y_i, \boldsymbol{X}_i^P, \boldsymbol{X}_i^W; W_i, H, T\right) \geqq 1\right\} \end{aligned} \quad (5.2)$$

$$\begin{aligned} D_i^R &= \left(Y_i, \boldsymbol{X}_i^P, \boldsymbol{X}_i^W; W_i, H, T\right) \\ &= \min_{P^P, P^W}\left\{\boldsymbol{P}_p X_i^P + \boldsymbol{P}_w \boldsymbol{X}_i^W : C_i^R = C_i^R\left(Y_i, \boldsymbol{P}_p, \boldsymbol{P}_w; W_i, H, T\right)\right\} \end{aligned} \quad (5.3)$$

The Lagrangian of the cost minimization problem in Eq. 5.3 is

$$\Lambda_i = \boldsymbol{P}_p \boldsymbol{X}_i^P + \boldsymbol{P}_w \boldsymbol{X}_i^W + \phi_i\left[1 - D_i^R\left(Y_i, \boldsymbol{X}_i^P, \boldsymbol{X}_i^W; W_i, H, T\right)\right] \quad (5.4)$$

Applying the envelope theorem to Eq. 5.4 gives:

$$\frac{\partial C_i^R}{\partial W_i} = \frac{\partial \Lambda_i}{\partial W_i} = -\phi_i \frac{\partial D_i^R}{\partial W_i} \quad (5.5)$$

$$\frac{\partial C_i^R}{\partial H} = \frac{\partial \Lambda_i}{\partial H} = -\phi_i \frac{\partial D_i^R}{\partial H} \quad (5.6)$$

In terms of input quantities, the first-order conditions are:

$$\frac{\partial C_i^R}{\partial X_i^P} = \frac{\partial \Lambda_i}{\partial X_i^P} = P_p - \phi_i \frac{\partial D_i^R}{\partial X_i^P} = 0 \quad (5.7)$$

$$\frac{\partial C_i^R}{\partial X_i^W} = \frac{\partial \Lambda_i}{\partial X_i^W} = P_w - \phi_i \frac{\partial D_i^R}{\partial X_i^W} = 0 \quad (5.8)$$

In terms of input prices, the first-order conditions are:

$$\frac{\partial C_i^R}{\partial P_p} = \frac{\partial \Lambda_i}{\partial P_p} = X_i^P \tag{5.9}$$

$$\frac{\partial C_i^R}{\partial P_w} = \frac{\partial \Lambda_i}{\partial P_w} = X_i^w \tag{5.10}$$

$$\frac{\partial C_i^R}{\partial \phi_i} = 1 - D_i^R(.) = 0 \tag{5.11}$$

Thus, at the optimum $\phi = \Lambda = \hat{C}_i^R$, where $\hat{C}_i^R(\ )$ is the minimum restricted cost (Shephard 1970). But it should be noted that $\hat{C}_i^R(\ )$ depends on the shadow prices that will be estimated. Therefore, in order to obtain $\hat{C}_i^R(\ )$ the assumption suggested by Färe and Grosskopf (1990, p. 125) that firms satisfy a balanced budget is adopted. Thus minimum restricted cost can be estimated since costs must equal revenues and when the distance function Eq. 5.1 is known, we can calculate the derivatives of the restricted cost function from the restricted distance function using:

$$-\left(\frac{\partial D_i^R}{\partial W_i}\right)\left(\hat{C}_i^R\right) = \frac{\partial C_i^R}{\partial W_i} \tag{5.12}$$

$$-\left(\frac{\partial D_i^R}{\partial H}\right)\left(\hat{C}_i^R\right) = \frac{\partial C_i^R}{\partial H} \tag{5.13}$$

From Eqs. 5.12 and 5.13, Proposition 1 can be stated: the accounting price of the groundwater stock of a renewable common property aquifer used for irrigated agriculture, corresponding to a symmetric perfect foresight open loop Nash equilibrium, is equal to the absolute shadow price of the resource derived from the restricted input distance function that describes firm-specific technology, or

$$-\left(\frac{\partial D_i^R}{\partial W_i}\right)\left(\hat{C}_i^R\right) = \mu_i \tag{5.14}$$

3 Econometric Specification, Empirical Estimation, and Results

Because all the farmers in our sample are located in the same (small) region, we do not observe any cross-sectional variation in input prices. For this reason, we can only estimate a production function (but not a cost function). Shadow groundwater scarcity rents on groundwater extraction costs are estimated using a stochastic restricted distance function using duality results between distance and cost functions. A translog stochastic input distance function for the case of K inputs and M outputs is estimated following Aigner et al. (1977). To obtain the frontier surface (i.e., the transformation function) we set Di = 1. Further, the restrictions required for homogeneity of degree +1 in inputs are imposed:

$$\sum_{k=1}^{K} \beta_k = 1; \sum_{l=1}^{K} \beta_{kl} = 0 \ \ for\, k = 1,2,\ldots,K \tag{5.15a}$$

and

$$\sum_{k=1}^{K} \beta_{km} = 0 \ for\, m = 1,2,\ldots,M \tag{5.15b}$$

Also the restrictions required for symmetry

$$\alpha_{mn} = \alpha_{nm} \quad for\, m,n = 1,2,\ldots,M \tag{5.16a}$$

and

$$\beta_{kl} = \beta_{lk} \quad for\, k,l = 1,2,\ldots,K \tag{5.16b}$$

And the conditions required for separability between inputs and outputs,

$$\delta_{km} = 0 \ for\ k = 1,2,\ldots,K \ and\ m = 1,2,\ldots,M \tag{5.17}$$

Thus, we have

$$= \alpha_0 + \sum_{m=1}^{M} \alpha_m lny_{mi} + \frac{1}{2}\sum_{m=1}^{M}\sum_{n=1}^{M} \alpha_{mn} lny_{mi} lny_{ni} + \sum_{k=1}^{K-1} \beta_k lnx^*_{ki} + \frac{1}{2}\sum_{k=1}^{K-1}\sum_{l=1}^{K-1} \beta_{kl} lnx^*_{ki} lnx^*_{li} + \sum_{l=1}^{K-1}\sum_{m=1}^{M} \delta_{km} lnx^*_{ki} lny_{mi} \tag{5.18}$$

$i = 1,2,\ldots,N \ and\ x^*_k = x_k / x_K$

Where i denotes the ith firm in the sample. It should be noted that homogeneity implies D(y, ωx) = ωD(y, x) for any ω > 0. One of the inputs was arbitrarily chosen and set $\omega = 1/x_K$. Therefore D(y, x/x_K) = D(y, x) /x_K. The frontier function has an error term with two components. The first component is a symmetric error term (V_i) that accounts for noise, which is assumed identically and independently distributed with zero mean and constant variance [$iid \sim N(0,\sigma_v^2)$]. The second component is an asymmetric error term (U_i) that accounts for technical inefficiency, which is assumed to follow an iid distribution truncated at zero (N(v, σ_U^2)). The two components of the error term, V_i and U_i, are independent. Predictions for Di = exp(Ui) are obtained using the conditional expectation Di = E[exp(Ui)|Ωi], where Ωi = Vi-Ui. Changing notation ln(Di) to Ui, Eq. 5.18 becomes:

$$-\ln\left(x_{Ki}\right) = TL\left(y_i, \frac{x_i}{x_{Ki}}, \alpha, \beta\right) + Vi - Ui \quad i = 1,2,\ldots,N \tag{5.19}$$

Equation 5.19 is estimated by maximum likelihood. Results are presented in Table 5.1. Data are drawn from a production surveys conducted in the agricultural region close to the Asopos River Basin in 2009. Farm-specific data includes: area of holding, land use and tenure, area planted, production of temporary and permanent crops, production inputs (including extracted groundwater), administrative costs, personal characteristics of buyers and sellers, employment of holders and family members, labor costs, value of construction works and other investments, indirect taxes and other expenses. The quality of the data-set is limited by the usual difficulties that one encounters when attempting to document inputs and outputs of agricultural activities. Particular difficulties where encountered in the collection of accurate groundwater extraction rates. The data-set has 301 cross sections. The following variables were used:

Output

y = firm-specific total value of output from production of agricultural crops, measured in Euros and deflated by the wholesale agricultural index.

Inputs

x_1 = farm-specific total area of non-irrigated land (acres),
x_2 = farm-specific annual labor costs (Euros),
x_3 = farm-specific total value of input costs, including fertilizers, manure, pesticides, fuel and electric power for groundwater extraction (Euros),
x_4 = farm-specific yearly groundwater extraction (m^3),
x_5 = farm-specific total area of irrigated land (acres); the negative of x_5 is the dependent variable of the estimated stochastic frontier.

Table 5.1 reports the estimated parameters for the Translog function. As it was expected, the inputs have positive signs and the outputs negative signs. The parameters for squared coefficients and interactions are also reported in the table. The estimated one-sided likelihood ratio (LR) suggests the rejection of the null hypothesis of no technical inefficiency. If the null hypothesis is truc, then the generalized LR statistic is asymptotically distributed as a mixture of chi-square distributions. The critical value for this mixed chi-square distribution is 2,706 for a 5 % level of significance (taken from Table 5.1 of Kodde and Palm [1986]). On the other hand, a value of one for γ ($\gamma = \sigma_U^2/(\sigma_U^2 + \sigma_V^2)$) indicates that all deviations are due to technical inefficiency, while a zero value indicates that the deviations from the frontier are entirely due to noise. It should be noted that γ is not equal to the ratio of the variance of the technical inefficiency effects to the total residual variance. Both hypotheses,

Table 5.1 Estimated parameters for the input distance function[a]

Variable	Parameter	ML estimates	Standard-error	T-ratio[b]
Constant	α_0	1.48E+00	1.03E+00	1.44E+00
Value of output	α_1	−3.96E-01	1.49E-01	−2.65E+00
Non irrigated land	β_1	6.14E-01	2.32E-01	2.65E+00
Labor costs	β_2	8.65E-02	7.26E-02	1.19E+00
Inputs costs	β_3	2.81E-02	1.15E-01	2.45E-01
Groundwater extraction	β_4	3.89E-01	1.04E-01	3.73E+00
0.5 Squared Value of Output	β_5	3.15E-02	1.21E-02	2.62E+00
0.5 Squared Non Irrigated land	β_6	−2.98E-01	3.49E-02	−8.55E+00
0.5 Squared Value of labor costs	β_7	2.79E-03	4.14E-03	6.75E-01
0.5 Squared inputs costs	β_8	−6.05E-03	1.42E-02	−4.25E-01
0.5 Squared groundwater extraction	β_9	−1.43E-01	8.60E-03	−1.66E+01
Value of output* Non irrigated land	β_{10}	1.43E-02	1.40E-02	1.02E+00
Value of output* Labor costs	β_{11}	−7.80E-03	4.84E-03	−1.61E+00
Value of output* Inputs costs	β_{12}	−2.97E-03	5.09E-03	−5.84E-01
Value of output* Groundwater extraction	β_{13}	−4.57E-03	6.00E-03	−7.62E-01
Non irrigated land* Labor costs	β_{14}	1.40E-02	6.82E-03	2.05E+00
Non irrigated land* Inputs costs	β_{15}	2.80E-02	1.06E-02	2.63E+00
Non irrigated land* Groundwater extraction	β_{16}	−7.19E-02	1.07E-02	−6.74E+00
Labor costs* Inputs costs	β_{17}	−6.56E-03	2.98E-03	−2.20E+00
Labor costs* Groundwater extraction	β_{18}	5.51E-03	3.05E-03	1.80E+00
Inputs costs* Groundwater extraction	β_{19}	2.03E-02	5.87E-03	3.47E+00
Irrigated land	log likelihood	−1.30E+02		
Irrigated land	LR test	4.94E+01		
Irrigated land	σ^2	3.55E-01	4.46E-02	7.96E+00
Irrigated land	Γ	8.91E-01	4.28E-02	2.08E+01

*Denotes a multiplication of two variables (i.e. a*b)
[a]The dependent variable is irrigated land. Number of cross sections is 301; number of time periods is 1
[b]Hypothesis tests are carried out at 95 % confidence level

$\gamma = 0$ and $\gamma = 1$, are rejected at the 95 % level of significance, supporting the existence of technical inefficiency and the choice of a stochastic model, respectively.

Firm-specific technical efficiencies are reported in Table 5.2. In this case, technical inefficiency means the use of an excessive amount of inputs to produce fixed output levels and could be related to the lack of incentives faced by the operators of the firm. In other words, the use of an economically suboptimal input mix denotes inefficiency in the allocation of resources. This could be the result of exogenous

Table 5.2 Predicted technical efficiency estimates[a]

Firm 1–50	Firm 51–100	Firm 101–150	Firm 151–200	Firm 201–250	Firm 251–301
8.38E-01	6.95E-01	7.18E-01	5.82E-01	7.07E-01	8.22E-01
7.52E-01	7.70E-01	7.85E-01	6.60E-01	7.38E-01	7.69E-01
9.50E-01	8.78E-01	7.85E-01	7.48E-01	7.55E-01	8.51E-01
7.84E-01	3.16E-01	6.48E-01	7.09E-01	7.27E-01	4.67E-01
8.28E-01	6.93E-01	7.28E-01	7.01E-01	5.04E-01	9.03E-01
8.87E-01	3.76E-01	7.23E-01	8.57E-01	6.42E-01	8.35E-01
7.50E-01	8.02E-01	7.59E-01	6.55E-01	7.49E-01	8.88E-01
8.18E-01	7.62E-01	7.10E-01	7.45E-01	7.18E-01	2.30E-01
8.41E-01	7.38E-01	7.85E-01	7.24E-01	7.14E-01	6.37E-01
6.23E-01	8.00E-01	7.41E-01	7.06E-01	7.10E-01	6.05E-01
7.43E-01	8.27E-01	7.07E-01	6.87E-01	5.76E-01	7.46E-01
8.59E-01	7.33E-01	8.23E-01	6.99E-01	8.04E-01	7.49E-01
4.84E-01	7.45E-01	7.73E-01	7.18E-01	6.80E-01	7.74E-01
7.60E-01	7.32E-01	7.46E-01	7.31E-01	6.60E-01	7.91E-01
9.05E-01	6.99E-01	7.70E-01	7.00E-01	2.69E-01	8.01E-01
6.94E-01	6.96E-01	6.48E-01	7.29E-01	2.48E-01	3.93E-01
7.45E-01	7.79E-01	6.49E-01	6.97E-01	2.56E-01	7.99E-01
8.61E-01	7.60E-01	7.48E-01	7.44E-01	7.37E-01	8.85E-01
8.79E-01	8.90E-01	6.89E-01	7.02E-01	7.40E-01	7.19E-01
8.48E-01	8.13E-01	7.15E-01	7.72E-01	6.91E-01	9.34E-01
7.08E-01	6.41E-01	6.94E-01	6.98E-01	6.91E-01	7.78E-01
7.01E-01	7.50E-01	8.27E-01	7.81E-01	9.00E-01	8.18E-01
7.37E-01	4.10E-01	6.74E-01	7.75E-01	6.78E-01	8.26E-01
4.94E-01	4.71E-01	6.54E-01	7.18E-01	7.24E-01	8.00E-01
4.98E-01	7.02E-01	2.07E-01	6.92E-01	7.23E-01	8.02E-01
2.86E-01	7.01E-01	7.19E-01	7.21E-01	7.72E-01	7.26E-01
7.28E-01	7.19E-01	2.80E-01	7.27E-01	8.56E-01	7.97E-01
5.91E-01	7.75E-01	7.21E-01	8.93E-01	7.25E-01	7.41E-01
6.28E-01	7.35E-01	8.15E-01	6.83E-01	6.71E-01	7.42E-01
7.43E-01	6.47E-01	3.35E-01	7.96E-01	7.17E-01	7.51E-01
7.60E-01	5.27E-01	2.55E-01	6.97E-01	7.08E-01	7.96E-01
2.00E-01	6.27E-01	9.10E-01	5.94E-01	8.02E-01	7.44E-01
7.04E-01	3.43E-01	4.39E-01	7.00E-01	7.02E-01	7.11E-01
6.23E-01	8.53E-01	6.88E-01	6.93E-01	6.87E-01	7.37E-01
4.34E-01	7.63E-01	5.64E-01	4.19E-01	3.58E-01	7.31E-01
8.95E-01	7.49E-01	7.18E-01	6.89E-01	4.75E-01	7.36E-01
7.38E-01	2.92E-01	6.99E-01	8.89E-01	9.47E-01	7.40E-01
8.03E-01	8.00E-01	4.82E-01	7.40E-01	4.60E-01	7.87E-01
7.29E-01	6.12E-01	7.30E-01	7.56E-01	6.95E-01	7.05E-01
7.18E-01	8.34E-01	3.57E-01	6.94E-01	8.62E-01	7.65E-01
6.17E-01	1.92E-01	7.53E-01	2.13E-02	3.71E-01	7.64E-01
6.99E-01	8.62E-01	7.19E-01	7.28E-01	7.83E-01	8.37E-01
7.64E-01	7.42E-01	7.45E-01	5.49E-01	7.41E-01	5.76E-01
5.51E-01	8.73E-01	7.10E-01	7.02E-01	9.21E-01	6.53E-01
7.24E-01	8.20E-01	7.14E-01	8.71E-01	9.37E-01	3.92E-01
6.74E-01	7.51E-01	6.66E-01	6.84E-01	7.17E-01	7.57E-01
6.24E-01	7.75E-01	7.88E-01	2.40E-01	8.33E-01	7.96E-01
4.98E-01	8.61E-01	6.77E-01	8.05E-01	8.90E-01	5.33E-01
7.20E-01	5.78E-01	8.89E-01	7.55E-01	8.85E-01	7.93E-01
7.24E-01	8.52E-01	6.97E-01	6.84E-01	5.00E-01	7.40E-01
					8.30E-01

[a]Mean efficiency is 0.69841938

environmental constraints. The existence of technical inefficiency alone does not necessarily imply biased cost function estimates. One use of these results is that these technical inefficiency measures can be used by the regulator for competitive benchmarking in which taxes or subsidies granted to each farm are based on the costs of a similar (in terms of input mix) but more efficient firm. Such a regulatory framework can:

1. Increase incentives for efficiency among the managers' of the farms and
2. Using these estimates, the informational asymmetry between the principal (regulators or consumers of agricultural products) and the agent (managers of the farms) can be reduced.

Finally, the mean per cubic meter shadow price estimates are calculated using Proposition 1. The mean annual per farm minimum restricted cost function $\hat{C}_i^R$ is approximated by the mean annual per farm revenue which is measured in Euros, 2005 constant prices: €651,951. The change in the restricted distance function per unit change in groundwater extraction $\left(\frac{\partial \ln D_i^R}{\partial \ln W_i}\right)$, measured in Euros per cubic meter, is the estimated parameter of the quantity of groundwater extraction from the stochastic distance function estimation: 0.389 €/m^3, the results of which are presented in Table 5.1. Moreover, D_i^R and W_i are respectively the mean annual estimated distance function (0.139 €/m^3) and mean groundwater extraction per farm (7,173 m^3). Finally, the estimated mean shadow value in situ of per cubic meter groundwater is $\mu_i = 0.154$ €/m^3.

4 Conclusions

Based on an input distance function an in situ shadow price was estimated in a framework independent of cost minimization restrictions. The estimated shadow price of in situ groundwater is €0.154 per cubic meter of water. The estimated groundwater's shadow price can be used to value the stock of water and to help in the design of pumping taxes in order to mitigate negative externalities (e.g. groundwater pollution and aquifer depletion) associated with groundwater use. The use of a distance function approach in estimating scarcity rents is supported by the existence of technical inefficiencies (as it evident from Table 5.2). Therefore, this approach is more appropriate than the restricted cost function approach. Further, these technical inefficiency measures can be used by the regulator. In this case, taxes or subsidies could be granted to each farm based on the costs of a similar (in terms of input mix) but more efficient firm. This kind of policy can increase the incentives towards efficiency, a challenging task when regulation of common property resources is done. Besides, this could reduce the information asymmetry between farmers, consumer and regulators, which is another major issue for the implementation of agricultural policies.

Appendix

Type of crops grown in the Asopos region

Crop	Number of farms	Total area (acres)	Share of total area (%)	Total irrigated area (acres)	Share of total irrigated area (%)
Temporary crops					
Barley	6	230	1.7	0	0.0
Beetroot	2	4	0.0	4	0.1
Carrot	1	15	0.1	15	0.3
Corn	1	20	0.1	0	0.0
Cotton	7	265	2.0	265	5.3
Crop	1	20	0.1	0	0.0
Cabbage	6	30	0.2	30	0.6
Oats	1	100	0.7	0	0.0
Okra	2	7	0.1	7	0.1
Onions	21	506	3.7	486	9.8
Potatoes	55	2,643	19.5	2,638	53.0
Tomatoes	1	200	1.5	200	4.0
Watermelon	1	6	0.0	6	0.1
Wheat	72	4,595	33.9	114	2.3
Beans	1	80	0.6	80	1.6
Cauliflower	1	5	0.0	5	0.1
Melon	1	70	0.5	70	1.4
Peas	1	10	0.1	10	0.2
Spinach	1	3	0.0	3	0.1
Permanent crops					
Grapes	2	50	0.4	50	1.0
Olives	125	3,712	27.4	202	4.1
Organic olives	2	19	0.1	0	0.0
Pistachios	4	160	1.2	160	3.2
Vineyard	25	822	6.1	632	12.7
Total		**13,572**	**100 %**	**4,977**	**100 %**

Wheat, olives, and potatoes are the three major crops grown in the region. They represent respectively 34 %, 27 %, and 20 % of the total cultivated area. About half of the total irrigated area is planted with potatoes (53 %). Vineyard and onions represent 13 % and 10 % of the total irrigated area in the sample, respectively.

Use of irrigation for each type of crop

Crop	Number of farms	Total area (acres)	Total irrigated area (acres)	Share of the area which is irrigated (%)
Temporary crops				
Barley	6	230	0	0
Beetroot	2	4	4	100
Carrot	1	15	15	100

(continued)

Appendix (continued)

Use of irrigation for each type of crop

Crop	Number of farms	Total area (acres)	Total irrigated area (acres)	Share of the area which is irrigated (%)
Corn	1	20	0	0
Cotton	7	265	265	100
Cabbage	6	30	30	100
Oats	1	100	0	0
Okra	2	7	7	100
Onions	21	506	486	96
Potatoes	55	2,643	2,638	100
Tomatoes	1	200	200	100
Watermelon	1	6	6	100
Wheat	72	4,595	114	2
Beans	1	80	80	100
Cauliflower	1	5	5	100
Melon	1	70	70	100
Peas	1	10	10	100
Spinach	1	3	3	100
Permanent crops				
Grapes	2	50	50	100
Olives	125	3,712	202	5
Organic olives	2	19	0	0
Pistachios	4	160	160	100
Vineyard	25	822	632	77
Total	**340**	**13,572**	**4,977**	**37 %**

Cereals (barley, corn, oats, wheat) are not irrigated in general. Only 5% of the area planted with olive trees is irrigated. Fields planted with cotton, fruits, and vegetables are fully irrigated. Overall, 37 % of the total area in the sample is irrigated. The three major products that are grown in Asopos are wheat, olives, and potatoes. We can see from this table that farmers do not combine wheat, olives, or potatoes with the growing of other products in most cases.

Crop	Farmers growing *wheat* also grow...	Farmers growing *olives* also grow...	Farmers growing *potatoes* also grow...
Barley	2	0	0
Beetroot	0	0	0
Carrot	0	0	0
Corn	0	0	1
Cotton	1	0	1
Crop	0	1	0
Cabbage	0	0	1
Oats	0	0	0
Okra	1	0	0

(continued)

Appendix (continued)

Crop	Farmers growing *wheat* also grow…	Farmers growing *olives* also grow…	Farmers growing *potatoes* also grow…
Onions	2	0	10
Potatoes	3	1	–
Tomatoes	0	0	0
Watermelon	0	0	0
Wheat	-	4	3
Beans	0	0	1
Cauliflower	0	0	0
Melon	0	0	1
Peas	0	0	0
Spinach	0	0	0
Grapes	0	0	0
Olives	4	–	1
Organic olives	0	0	0
Pistachios	0	0	0
Vineyard	0	1	2

The three major crops in the area are wheat, potatoes and olives, which we will consider in turn.

4.1 Wheat Producers

In what follows we consider the 59 farmers who grow only wheat (overall 72 farmers grow wheat in our sample). The following inputs are considered: fertilizers, pesticides and labor. Fertilizers and pesticides use are farmers' statements while labor is calculated as follows: number of days of casual workers + number of permanent workers × 250. Some basic statistics are shown below. There are all on a *per acre* basis.

Variable	Obs.	Mean	Std. Dev.	Min	Max
Production (tonnes/acre)	59	0.27	0.19	0.02	0.80
Fertilizer use (kg/acre)	58	17.07	15.37	0.00	60.00
Pesticides use (kg/acre)	55	0.27	0.60	0.00	2.14
Labor use (days/acre)	59	0.15	0.27	0.00	1.50
Land (acre)	59	70.61	85.86	6.00	500.00

All these statistics are on a *per acre* basis so the figures should not vary too much from one farmer to the other. However we observe very large variations. For example, fertilizer use varies from 0 kg/acre to 60 kg/acre, with a mean of 17 kg/

acre. The farmers stating 0 use of fertilizer, pesticides or labor probably did not want to answer or did not know. For these farmers, I have replaced 0 by the median value in the sample of farmers growing wheat only.

4.2 *Statistics on Yield*

1 *Acre* – US, = 0.4046873 *ha*, 1 *hectare* (*ha*), = 2.471044 *acre* (US)

Farmers in the sample produce on average 0.27 tonnes per acre, which corresponds to 0.67 tonnes per hectare (or 670 kg per ha). The average wheat yield in Greece is 1,900–3,000 kg/ha. The average yield on the sample thus seems a bit low. Once all variables are transformed in logs a Cobb Douglas production function is estimated. Because of the small sample size, it is not reasonable to estimate a Translog production function.

OLS estimation results – Cobb Douglas production function (59 obs)

	Coef.	Std. Err.	P>t
Fertilizer	0.026	0.093	0.778
Pesticides	0.204	0.119	0.092
Labor	0.140	0.080	0.087
Constant	−0.127	0.146	0.386

In this model the dependent variable is wheat yield. The explanatory factors are the three inputs measured in physical terms: fertilizer use per acre, pesticides use per acre, labor use per acre. The three estimated coefficients have the expected positive sign but only two are significant at the 10 % level. However, the model is not significant overall (p-value of the Fisher test is 0.1251). As a consequence the adjusted R-squared is also quite low: 0.0491.

4.3 *Potatoes Producers*

In what follows we consider the 34 farmers who grow only potatoes (overall 55 farmers grow potatoes in our sample). Some basic statistics are shown below.

Variable	Obs	Mean	Std. Dev.	Min	Max
Production (tonnes/acre)	34	2.04	1.13	0.20	5.00
Fertilizer use (kg/acre)	34	20.90	19.71	0.23	75.30
Pesticides use (kg/acre)	25	0.94	1.18	0.00	4.00
Labor use (days/acre)	34	0.80	1.28	0.00	6.50
Water use (m3/acre)	33	2.02	11.61	0.00	66.67
Land (acre)	34	52.15	37.60	7.00	150.00

Here too some figures are really surprising: fertilizer use varies from a low of 0.23 kg/acre to a high of 75.30 kg/acre. Again the zeroes for pesticides and labor do not make much sense.

4.4 Olive Producers

In what follows we consider the 117 farmers who grow only olives (overall 125 farmers grow olives in our sample). Some basic statistics are shown below.

Variable	Obs.	Mean	Std. Dev.	Min	Max
Production (tonnes/acre)	104	0.14	0.18	0.02	1.13
Fertilizer use (kg/acre)	112	20.42	18.55	0.00	80.00
Pesticides use (kg/acre)	107	0.19	0.84	0.00	6.00
Labor use (days/acre)	117	2.99	23.20	0.00	250.38
Land (acre)	117	29.91	23.84	5.00	120.00

References

Aigner, D., Lovell, C. A. K., & Schmidt, P. J. (1977). Formulation and estimation of stochastic frontier production function models. *Journal of Econometrics, 6*(1), 21–37.

Coelli, T., & Perelman, S. (2000). Technical efficiency of European railways: a distance function approach. *Applied Economics, 32*(15), 1967–1976.

Färe, R., & Grosskopf, S. (1990). A distance function approach to measuring price efficiency. *Journal of Public Economics, 43*, 123–126.

Färe, R., Grosskopf, S., Lovell, C. A. K., & Yaisawarng, S. (1993). Derivation of shadow prices for undesirable outputs: A distance function approach. *The Review of Economics and Statistics, 75*(2), 374–380.

Färe, R., Grosskopf, S., & Lovell, C. A. K. (1994). *Production frontiers*. New York: Cambridge University Press.

Färe, R., & Primont, D. (1995). *Multi-output production and duality theory and applications*. Norwell: Kluwer.

Grosskopf, S., & Hayes, K. (1993). Local public sector bureaucrats and their input choices. *Journal of Urban Economics, 33*, 151–166.

Howe, C. (2002). Policy issues and institutional impediments in the management of ground water: Lessons from case studies. *Environment and Development Economics, 7*, 625–642.

Koundouri, P., & Xepapadeas, A. (2004). Estimating accounting prices for common pool natural resources: A distance function approach. *Water Resources Research, 40*(6), W06S17.

Shephard, R. W. (1970). *Theory of cost and production functions*. Princeton: Princeton University Press.

Chapter 6
A Choice Experiment for the Estimation of the Economic Value of the River Ecosystem: Management Policies for Sustaining NATURA (2000) Species and the Coastal Environment

P. Koundouri, R. Scarpa, and M. Stithou

Abstract The valuation method of Choice Experiments (CEs) is often used for the economic valuation of natural areas with several nonmarket features that are either degraded or under-degradation. This method can be used to obtain estimates of Willingness-to-Pay (WTP) for the sustainability of several features of natural ecosystems. In particular, the CE method is a survey-based nonmarket valuation technique which can be used to estimate the total economic value of an environmental good in the form of a stock or a service flow as well as the value of its component attributes. Particularly, the bundle of improvements that have been valued in the Asopos water catchment and presented in this chapter is a mixture of use and non-use values. These include: (*a*) environmental conditions described in terms of ecological status in all water bodies of the catchment, (*b*) impact on the local economy in terms of tourism/recreation, demand for local production and cost of living for households and (*c*) impact on human health described as availability of water with a quality and quantity sufficient for satisfying different local uses. It should be also noted that the survey has been administered in samples of respondents from both the Asopos catchment area (more rural) and the Athens area (more urban), since there is the belief that residents of the Asopos River Basin (RB) are not the only ones who would benefit from the environmental improvements taking place in Asopos area. From a broader policy perspective the goal is to derive estimates of values to inform a cost-effectiveness analysis for the determination of the optimal program of measures as suggested in the content of Article 11 of Water Framework Directive (WFD).

P. Koundouri (✉) • M. Stithou
Athens University of Economics and Business, Athens, Greece
e-mail: pkoundouri@aueb.gr; stithoumavra@yahoo.com

R. Scarpa
Waikato Management School, Hamilton, New Zealand
e-mail: rscarpa@waikato.ac.nz

P. Koundouri and N.A. Papandreou (eds.), *Water Resources Management Sustaining Socio-Economic Welfare*, Global Issues in Water Policy 7,
DOI 10.1007/978-94-007-7636-4_6, © Springer Science+Business Media Dordrecht 2014

Keywords Non market valuation • Choice experiments • Asopos River Basin • Environmental degradation • Water quality and quantity

1 Introduction

As described in Chap. 1 the river Asopos runs across the Eastern river basin district of Greece, which is located about 60 km north of Athens and it is one of the 14 water basin districts of the country. The catchment of the river Asopos covers an area of 724 km^2 and it is one of the three catchments of the Eastern river basin district. Chapter 2 of this volume offered a description of the socio-economic characteristics of current developmental pressures on the Asopos RB.

In this chapter it was discussed how the Asopos hydrometric area bears witness to the degradation which was mostly caused by unregulated human activity and took place despite the obviously high ecological value of the basin area. The Asopos River is a tributary to a sea-water lake of significant ecological importance which hosts rare habitats for protected fauna and for transient populations of migrating birds that use it as a temporary resort during their migration to and from distant locations. Therefore, in the Asopos estuary and in the nearby wetland of Oropos, which is the second most important wetland of Attiki, bird wildlife is particularly important. The wetland is to support a habitat which hosts an estimated number of more than 140 bird species, many of which are protected by EU legislation,[1] and they comprise raptors, herons, waterfowl present during the winter (November to February), waders all year round, with their peak during migration in winter. Important is considered the presence of *Larus melanocephalus* that is endangered according to the Red Data Book, among others. What makes the whole of this area an important migratory passage is its relative scarcity and difficult substitutability, since it is one of the few remaining wetlands of Attiki. In addition, the areas by the sea-water lake and the nearby coastal zone represent a significant tourist attraction where different recreational activities take place. As reported in Chap. 4 the tourist area of Asopos RB is mainly consisted of vacation or second residences. The number of vacation or second residences is 16,267 and is the 37.6 % of the total residences in Asopos RB. It is worth noting that from the total of second or vacation residences most of them are located in the coastal areas of the Municipalities. A further impact is the decreased demand for the local agricultural production and the high cost of living that the households face as a consequence of not having access to clean water which have seriously affected the local economy.

The purpose of this chapter is to report the results of a multi-attribute stated preference choice survey specifically designed to estimate the use and non-use values that the two sub-populations of residents in Asopos and Athens hold for the general improvements that can be brought about in the Asopos catchment. Hence, the

[1] 31 are listed in Annex I of Directive 79/409/EEC(1) while in the area the following reproduce: Charadrius alexandrinus, Himantopus himantopus, Sterna albifrons and Calandrella brachydactyla.

non-market valuation method of CEs was used to estimate the WTP for changes in selected attributes (environmental conditions, impact on the economy and changes in potential water uses) consistent with the major problems identified above under alternative future scenarios.

Apart from the need to evaluate the socio-economic and environmental impacts related to the basin's degradation, this study also aimed to explore how the two different populations, the rural population resident in Asopos and the urban population resident in Athens, value the same set of proposed improvements. Another reason apart from the socio-demographic composition that has motivated this sampling is the different way those populations experience the environmental degradation due to location and economic dependence on the area.

Residents of Asopos and of Athens differ in various ways making optimal policy design that can satisfy the needs of both populations challenging. Athens is the national capital city, with a population of three million while Asopos basin has a much smaller permanent population of only 70,575. In addition, the populations differ in their demographic composition. For example, in Asopos 24 % of the resident population aged 15 or over is employed in the primary and secondary sector while only 18 % in the tertiary sector. In Athens these proportions change by a factor of 2, with 11 % of the equivalent aged population occupied in the primary and secondary sectors while 32 % in the tertiary sector. In addition, different average educational attainments are expected in the two populations.

The allegedly polluting industrial facilities located in the area crossed by the Asopos River have been targeted by local people as the source of major environmental degradation and health problems. However, the economy of the local community is highly dependent on the functioning of such facilities to support the local industrial activity. Nevertheless, a big part of the population feels that the cost borne by the local community is intolerably high when compared with the benefits produced by the local industrial activity.

Up until 2004, drinking water qualitative analyses were limited to microbiological and chemical controls. At the same time, a local mobilization started with information reaching EU through reports and complaints and written questions by members of the European Parliament. Local people pressured the government to conduct more detailed chemical analysis because of fear of the water being polluted by the highly toxic hexavalent chromium. The rigidity of the government and local authorities to provide any official response, induced mistrust in the local population towards the so-called "expert knowledge" behind the official positions held on the issue by the government and by private companies. This eventually led to a widespread mistrust on the scientific validity of the technical reports which tended to underplay the severity of the water pollution problem (Passali 2009). The government tended, at least at first, to deny the severity of the problem and appeared unwilling to embrace legal action against polluters and enforce adequate legislation. This in turn, reinforced public's mistrust. Nowadays the Asopos River anti-pollution movement has attracted the attention of several media raising public awareness about the problem. As a result, the country has become aware of the actual dimension of the pollution problem in the area sympathizing with the resident population.

Considering the socio-economic differences, the scale of the problem, the populations' personal experiences, their economic dependence on the area as well as the political implications associated with the management of the Asopos RB, makes it interesting to explore how the two populations form their values related to the same set of proposed categories of improvement.

2 Data and Survey Design and Description

Because the study's objective was to investigate use and non-use values, the target population was defined as the residents in the survey location, the Asopos basin, as they will be affected by changes in water management as well the residents of Athens which is in close proximity to the basin. Interviews took place in households and one adult per house participated. Quota sampling was followed according to 2001 Census data in order for the samples to be as representative as possible of the targeted population and every forth residence was called.

The survey design followed the recommended five steps for the conduct of a CE survey for the purpose of non-market valuation. That is, the selection of attributes, the definition of attribute levels, the choice of the experimental design to allocate alternative scenarios to choice tasks to present to respondents, and the elicitation of preferences by asking respondents to rank the alternative scenarios in each choice task.

As it became evident from the extensive analysis presented in the early chapters of the book the main impacts of the degradation of the area affect the environment, the local economy and human health. As a result, when selecting the CE attributes these dimensions were those primarily considered in the development of the scenario descriptions to be used in the survey design. The attributes and their levels associated with different management options are presented in Table 6.1.

Suitable showcards to depict each alternative scenario with pictures were prepared, each accompanied by simple descriptions that were read aloud during the questionnaire administration, which was administered by door-to-door interviews. So, this allowed enumerators to better illustrate policy outcomes to respondents in terms of attributes and levels. In particular, one showcard described the environmental condition of Asopos RB explaining the three water quality levels named as "Bad", "Moderate" and "Good" with regard to the river, Oropos lagoon, coastline, and groundwater, respectively. The second card described in three levels (Negative, Improved by 2015, Positive by 2027) the situation of the local economy in terms of tourism/recreation, demand for local products and cost of living for households. Other material used was a map of the RB showing its geography.

All non-monetary attributes had two levels of policy action, while a third level was associated with the no action option (Option A) (Fig. 6.1), which was used as the *status quo* outcome. This alternative was included in all choice sets and represented the outcome of no intervention and hence no cost to the residents. Five levels of cost were used. The payment vehicle proposed to respondents was an increase in the water bill to be paid per household and per year for the next 15 years. In addition

Table 6.1 Attributes and levels

Attribute	Some policy action/improvement	*Status Quo* (option A)	Variable name
Environmental condition	Moderate, Good	Bad	
Impact on local economy	Improved by 2015, Positive by 2027	Negative today	
Human health	Water suitable for all uses (drinking, cooking and irrigation), water suitable for some uses (drinking and cooking)	Water not suitable for drinking, cooking and irrigation	
Cost € (Tri-monthly water bill per household for the next 15 years)	2,4,6,8,12	0	

	Option A	Option B	Option C	Option D	Option E
Environmental condition	Bad	Bad	Good	Bad	Moderate
Impact on local economy	Negative today	Improved by 2015	Positive by 2027	Improved by 2015	Negative today
Human health	~~Drinking~~ ~~Cooking~~ ~~Irrigation~~	**Drinking Cooking Irrigation**	~~Drinking~~ ~~Cooking~~ ~~Irrigation~~	**Drinking Cooking Irrigation**	~~Drinking~~ ~~Cooking~~ ~~Irrigation~~
Cost (Tri-monthly water bill per household for the next 15 years)	0	4	6	6	4
1. Which option do you prefer most?	☐	☐	☐	☐	☐
2. Which option do you prefer least?	☐	☐	☐	☐	☐
3. Which one from the three remaining do you prefer most?	☐	☐	☐	☐	☐
4. Which one from the two remaining do you prefer least?	☐	☐	☐	☐	☐

Fig. 6.1 Example of a choice card

it was stated that money will be collected in a fund run by an organization specifically established for the improvement and conservation of Asopos' catchment, while an independent body such as EU will assure that money will be spent for that purpose. A preliminary pilot study of 30 randomly selected residents in each site (Asopos and Athens) was carried out to test the questionnaire and collect the priors to be used in the experimental design for the final survey. For the priors the employed design had 6 blocks of 12 choice cards.

The method used to extract from the full factorial the 36 choice tasks used in the experiment was based on the minimization of the expected D-error. The D-error is the determinant of the asymptotic variance-covariance matrix of the multinomial logit model (Sándor and Wedel 2001; Ferrini and Scarpa 2007; Rose and Bliemer 2009). For the standard multinomial logit the D-error is a function of the design matrix and of the values of the utility coefficients and the specification of the model, but not of the dependent variable defining choice. Starting from assumptions (Bayesian priors) on the values of the utility coefficients, as derived from the pilot study, and on their distributions (in our case multivariate normal) we computed this expectation and run a search to minimize the value of its determinant over the space of the design matrix values. This was executed using the Ngene Software, which is specialized software for stated choice design. Importantly, the design was obtained by assuming an indirect utility dummy coded, so that the design variables were not expressed on the levels.

The response task was framed as a sequential choice process, with respondents instructed to choose the most preferred alternative out of the initial 5 alternatives in the choice set. This best alternative was then excluded from the choice set. Then they were asked to select the least preferred out of the remaining 4, which was also excluded. This process was repeated for the remaining three alternatives from which the respondent selected the second most preferred out of the remaining 3, and finally the second least preferred out of 2. This approach is called the "repeated best-worst" approach and gives rise to a full preference ranking of the five alternatives in the choice set. Focusing on the preference extremes (best and worst) is believed to be less cognitively taxing for the respondent than alternative approaches (Scarpa et al. 2011).

Data for the final survey were collected from September to October 2011 by trained interviewers. The final useable samples collected consisted of 150 respondents from Asopos RB and 150 from Athens. The average completion time for an interview across those who completed the ranking tasks was 30 min.

3 Method

As stated before the objective of the study is that of estimating marginal WTP for different attribute and attribute levels as described in the scenarios presented to respondents for the two sub-populations of Athens and Asopos. In the analyses presented here we used only choices from the first and second best, rather than using the whole set of repeated best/worst observations. Selection of the best option gives rise to random utility maximization consistent logit probabilities (McFadden 1974), while selection of worst alternatives does not. So, the pseudo-choice sets used here were 24 for each respondent, 12 pseudo-choice sets provided by the best selections from 5 alternatives (first round of bests) and the other 12 pseudo-choice sets by the best selections from 3 alternatives (second round of bests).

We used a piece-wise linear coding to capture the effects on utility of increasingly larger improvements on the three attributes of interest. Piece-wise coding allows the analyst to estimate coefficients related to marginal improvements on a scale of monotonic changes. Attribute levels here offered two sets of gradual improvements. The coding for a 2 step improvement, as the one adopted here, is as follows: baseline (0,0), first level of improvement (1,0), maximum improvement (1,1). As can be seen the maximum improvement recognizes that a previous first level of improvement was already in place by maintaining a value of 1 in the first column. As a consequence the value of the beta coefficient associated with this extreme improvement captures only the further effect beyond that captured by the beta of the first level of improvement. This is in contrast with the standard dummy variable approach in which the coding of the maximum improvement would be (0,1), and therefore the utility coefficient would capture the jump in utility from the baseline. Piece-wise coding imposes consistency with weak monotonicity across coefficient estimates, while dummy coding does not. For example, consider the levels of the attribute Environmental Condition which are "Bad" (the current level), "Moderate" (a potential future level, in our language the first level of improvement) and "Good" (the extreme improvement level). Because the first level is in common between the coding of the "Moderate" and "Good" the coefficient on "Good" will capture only the utility effect that this further improvement produces beyond that captured in a "moderate" improvement.

In terms of the specification used, given our objectives, we estimated a linear indirect utility function for each of the two subsamples. We do this in two different selections of the sub-samples. The first includes all respondents, while the second excludes those that displayed a serial non-participation choice behavior. That is, all those that consistently chose the status-quo condition across all 12 first best decisions on the full set choice tasks. For these respondents the alternative to the status-quo offering various improved scenarios were never sufficiently appealing to motivate a payment. So, we have two models, one estimated on a sample *with* and another *without* serial non-participants, for each of the two sub-populations of beneficiaries (Athens and Asopos). Estimates of WTPs for the different levels of policy attributes are then derived from each of these models. Finally, we look at overlapping confidence intervals across marginal WTP estimates between the two samples to see which of these are statistically different from each other.

4 Results and Discussion

The results of the analyses show that in both sub-samples the exclusion of serial non-participants produces an increase in the model fit and are reported in Table 6.2 for Athens and Table 6.3 for Asopos (samples with all respondents) and in Table 6.4 for

Table 6.2 Model for Athens (all respondents)

	RUM estimates			Marginal WTP estimates					
	Coeff.	St.err.	|z-value|	Coeff.	St.err.	|z-value|	p-value	95% conf. int.	
Status Quo ASC	0.840***	0.090	9.3	7.28***	0.97	7.5	<0.01	5.38	9.18
Env. moderate	1.162***	0.061	18.9	10.07***	0.70	14.3	<0.01	8.69	11.44
Env. good	0.279***	0.057	4.9	2.41***	0.52	4.7	<0.01	1.40	3.43
LocalEcon Improved2015	0.465***	0.053	8.8	4.03***	0.48	8.4	<0.01	3.09	4.97
LocalEcon Positive2027	−0.206***	0.054	−3.8	−1.78***	0.46	−3.9	<0.01	−2.69	−0.88
Water for some uses	0.655***	0.062	10.5	5.68***	0.60	9.5	<0.01	4.50	6.85
Water for all uses	0.723***	0.063	11.4	6.27***	0.67	9.3	<0.01	4.95	7.59
Cost	−0.115***	0.007	−16.7						

Log likelihood function at Max. −4178.74 Inf.Cr.AIC = 8373.5 AIC/N = 2.326
* $p<.1$; ** $p<.05$; *** $p<.01$

Table 6.3 Model for Asopos (all respondents)

	RUM estimates			Marginal WTP estimates					
	Coeff.	St.err.	|z-value|	Coeff.	St.err.	|z-value|	p-value	95% conf. int.	
Status Quo ASC	1.071***	0.089	12.0	8.31***	0.91	9.2	<0.01	6.53	10.09
Env. moderate	1.237***	0.063	19.7	9.59***	0.63	15.3	<0.01	8.37	10.82
Env. good	0.060	0.057	1.1	0.47	0.45	1.1	0.294	−0.41	1.34
LocalEcon Improved2015	0.220***	0.053	4.1	1.70***	0.41	4.1	<0.01	0.90	2.51
LocalEcon Positive2027	−0.145***	0.055	2.6	−1.13***	0.42	2.7	0.008	−1.96	−0.29
Water for some uses	0.941***	0.066	14.3	7.29***	0.60	12.2	<0.01	6.12	8.47
Water for all uses	0.665***	0.064	10.5	5.16***	0.58	8.9	<0.01	4.03	6.29
Cost	−0.129***	0.007	18.2						

Log likelihood function −4133.05 Inf.Cr.AIC = 8282.1 AIC/N = 2.301
* $p<.1$; ** $p<.05$; *** $p<.01$

Athens and Table 6.5 for Asopos (samples without serial non-participants).[2] This is shown by looking at the average AIC value that decreases from 2.326 to 2.095 in the Athens sub-sample and from 2.301 to 1.961 in the Asopos one. Furthermore, the ASC for the *status quo* is never significant in the models without serial non-participants. This indicates that when these respondents are removed there is no systematic effect to stay with the current condition and avoid the proposed alternative scenarios.

The cost coefficient is negative and significant in all models, allowing the derivation of welfare estimates. While our expectation was to find all positive signs in the utility coefficients, we note that for the extreme improvement of the local

[2] A broad prior for ENV_2 was used. As a result, the same priors for the means of the parameters were used except for ENV_2, here we assumed a uniform distribution with a large range.

Table 6.4 Model for Athens (serial non participants excluded)

	RUM estimates			Marginal WTP estimates					
	Coeff.	St.err.	|z-value|	Coeff.	St.err.	|z-value|	p-value	95% conf. int.	
Status Quo ASC	−0.174	0.119	−1.5	−1.629	1.098	−1.5	0.138	−3.78	0.52
Env. moderate	1.181***	0.068	17.3	11.070***	0.852	13.0	<0.01	9.40	12.74
Env. good	0.414***	0.064	6.5	3.881***	0.679	5.7	<0.01	2.55	5.21
LocalEcon Improved2015	0.573***	0.058	9.9	5.367***	0.613	8.8	<0.01	4.17	6.57
LocalEcon Positive2027	−0.227***	0.059	−3.9	−2.124***	0.54	−3.9	<0.01	−3.18	−1.07
Water for some uses	0.621***	0.068	9.1	5.824***	0.696	8.4	<0.01	4.46	7.19
Water for all uses	0.954***	0.071	13.4	8.941***	0.949	9.4	<0.01	7.08	10.80
Cost	−0.107***	0.008	−14.2						

Log likelihood function −3134.40 Inf.Cr.AIC = 6284.8 AIC/N = 2.095
* $p<.1$; ** $p<.05$; *** $p<.01$

Table 6.5 Model for Asopos (serial non participants excluded)

	RUM estimates			Marginal WTP estimates					
	Coeff.	St.err.	|z-value|	Coeff.	St.err.	|z-value|	p-value	95% conf. int.	
Status Quo ASC	−0.180	0.128	−1.4	−1.423	1.006	−1.4	0.157	−3.39	0.55
Env. moderate	1.448***	0.075	19.2	11.476***	0.794	14.5	<0.01	9.92	13.03
Env. good	0.124*	0.068	1.8	0.982*	0.549	1.8	0.074	−0.09	2.06
LocalEcon Improved2015	0.337***	0.062	5.5	2.671***	0.495	5.4	<0.01	1.70	3.64
LocalEcon Positive2027	−0.203***	0.064	−3.2	−1.606***	0.494	−3.3	0.001	−2.57	−0.64
Water for some uses	1.052***	0.076	13.9	8.337***	0.704	11.8	<0.01	6.96	9.72
Water for all uses	1.001***	0.077	13.1	7.930***	0.824	9.6	<0.01	6.32	9.55
Cost	−0.126***	0.008	−15.5						

Log likelihood function −2698.42 Inf.Cr.AIC = 5412.8 AIC/N = 1.961
* $p<.1$; ** $p<.05$; *** $p<.01$

economy (positive in 2027) we obtain a negative and significant sign. What this sign tells us is that with respect to the utility impact of the intermediate improvement (0.465) an additional improvement is −0.206 utility unit smaller, which means that with respect to zero it still produces a positive improvement of 0.465–0.206 = 0.259 utility units.

Because the *z*-value significance is defined with respect to a null of zero, whereas the benchmark for a further notch in the scale of gradual improvements is in fact the value of the immediately previous improvement, we note that a test should be run with respect to the lower extreme of a confidence interval around the coefficient estimate for the local economy (improved 2015). Such lower extreme in the case of the model for Athens with all respondents is $0.465–1.96\times0.053 = 0.361$. Computing the cumulative probability at 0.361 for a random variable distributed $\Phi(\mu=0.259, \sigma=0.054)$ we find that it falls to the left of this value with very high probability. We conclude that the additional marginal effect for the 2027 scenario is actually valued less than what is the case for the proposed 2015 scenario. Maybe this is so because it was too far away in time for most respondents to be able to relate to it or perhaps because it was not clear in most respondents' mind that it implied the 2015 target. This result is found only for this attribute level and it is consistently negative across all models.

Turning the attention to the other two attributes, it is found that their utility effects jump up significantly in the moderate improvement levels, and also for the extreme improvements. The estimated marginal WTP effects are reported to the right columns of the tables and for the extreme improvement are to be interpreted as additional effects over the WTP for the first improvement levels. So, for example, the estimated marginal effect of €2.41 for "Env. Good" is to be interpreted as €2.41 over the €10.07 effect of "Env. Moderate". However, it is easy to see that when accounting for the variability implied by the approximate standard error of these estimate (obtained with the delta method) the upper extreme of a 95 % confidence interval for the latter is $10.07+1.96\times0.7 = 11.44$ while the lower extreme for the $(10.07+2.41)-1.96\times0.52 = 11.46$, which is just a bit to the right of €11.44. This leads us to conclude that the marginal effect of moving from moderate to good in this model is estimated to be very low. In the model without serial non-participants this effect is not much higher. A similar conclusion is to be reached in the Asopos estimations.

On the other hand, the estimated marginal effect of extending water uses from "some uses" to "all uses" emerges as adding substantial value, nearly twice as much as the first level of improvement. This result is consistent across all models, and unsurprisingly, it reaches the level of "some uses" to be valued more by the local residents of Asopos (€7.29 in the sample with all respondents and €8.33 in that without serial non participants) than it is by those of Athens (€5.68 in the sample with all respondents and €5.82 in that without serial non participants). However, the Athenians show a higher WTP for the further step in improvement with €6.27 in the sample with all respondents versus €5.16 by the Asopos analogue sub-sample; and €8.94 in the Athens sub-sample without serial non participants versus €7.93 by the Asopos analogue sub-sample.

5 Conclusions

The water catchment of the Asopos River has witnessed a severe decline of environmental quality over the recent past. There is a clear interest to improve such conditions from both the resident rural population and the population of urban dweller that makes most use of it as recreationists.

The elicited monetary values from the set of choice collected in the survey demonstrate the importance of the improvements for both the residents of Asopos and Athens. Furthermore, it is regarded that these monetary values are policy relevant and in the absence of anything better they can be employed in order to assess the effectiveness of incurring the cost recovery of water resources in the area considering also the environmental cost involved. Cost effectiveness analysis for the determination of the optimal program of measures was suggested by Article 11 of WFD (CEC 2000) and it represent the core principle driving public investment, especially in harden economic times as at present.

Aggregating results considering the models with all the respondents for the case for example of improvements in water available for all uses, we can see that in the case of Asopos the WTP is about €475, 000 per year (for the years 2012–2027) (23,000 households are permanently connected with the public system of water supply in Asopos RB) while for the Athens' sample is €45,144,000 per year (1,800,000 households). Moderate environmental improvements are valued at €882,200 per year in Asopos and at €72,500,000 per year in Athens.

Other evidence about the economic damage in Asopos basin is offered by the application of a Contingent Valuation Method (CVM) (Dimaras et al. 2010) aimed to elicit WTP of the catchment's residents for improvements in the area's groundwater. From a sample of 154 revealed that on average the households of the area were willing to pay €400 per year for the next 10 years to an independent management body entrusted with the remediation of the polluted groundwater within the 10 years period. Aggregating to the number of households that were willing to contribute to the voluntary scheme at a 3 % discount rate resulted in 1M € annually (about €8.5M in 10 years). In the same study it is reported that the construction of a new pipeline connection to provide Oropos area with clean water is estimated at €9,400,000.

Similarly, Papadiochou et al. (2011) elicited the cost of the environmental damage based on Attica households' willingness to pay a yearly contribution, in voluntary basis, to a new organization that will take measures and will remediate groundwater pollution in the next 10 years. CVM answers were collected from a sample of 400 households by telephone interviews contacted randomly. Households were willing to pay, based on the lower bound average WTP, an annual contribution of €45 approximately in order to support the organization. Taking into account the population of interest, the annual aggregated value is about €60 M, which corresponds to a present value of €470 M (annual payments for 10 years, discounted at real interest rate of 5 %). It is also noted that households' WTP covers the total of the services provided by the aquifer in question, including non-use values such as the protection of the function of ecosystem services.

Another study that focused on the estimation of the economic damage of groundwater degradation is that of Laoudi et al. (2011). The authors note that the least-cost approach for pollution abatement measures regarding public water supply would comprise of replacing groundwater with surface water for Inofyta area and of installing a central Reverse Osmosis water treatment system for Oropos area. The authors argue that the total cost of these measures estimated at €426K annually is considered a solid basis for valuing the economic damage by groundwater pollution in the area due to the loss of residential use of groundwater.

Finally, Loizidou's (2009) study suggested the construction of a Central Wastewater Processing Unit which would gave the possibility to the industries and the Municipality of Avlonas to dispose their wastes. The total construction cost of the Central Unit was estimated at €33M with the construction cost of network connection at €14–15M. In addition, the annual running cost for an average daily provision of 16,762 m^3 was estimated as $0.611*365*16{,}762 \approx$ €3,738,177. Overall, the reported estimates in this chapter provide evidence of the importance of the Asopos RB for residents in both Asopos and Athens who are willing to contribute a considerable amount towards the cost of necessary mitigation measures.

References

Commission of the European Communities (CEC). (2000). *Directive of the European Parliament and of the Council establishing a framework for Community action in the field of water policy.* 1997/0067 (COD), C5-0347/00.

Council Directive 79/409/EEC of 2 April 1979 on the conservation of wild birds.

Dimaras, A., Mastrogiannis, F., & Damigos, D. (2010). *Estimation of the cost of pollution of Asopos River.* MSc dissertation, National Technical University of Athens, School of Mining and Metallurgical Engineering. (In Greek)

Ferrini, S., & Scarpa, R. (2007). Designs with a-priori information for nonmarket valuation with choice-experiments: a Monte Carlo study. *Journal of Environmental Economics and Management, 53*(3), 342–363.

Laoudi, A., Tentes, G., & Damigos, D. (2011). *Groundwater damage: A cost-based valuation for Asopos River basin.* Proceedings of the 3rd International CEMEPE & SECOTOX Conference Skiathos, June 19–24, 2011, ISBN 978-960-6865-43-5.

Loizidou, M. (2009). *Environmental impact assessment for a central processing unit for the industrial wastewater of Asopos area and the urban wastewater of the municipality of Avlonas.* National Technical University of Athens, School of Chemical Engineering. (In Greek)

McFadden, D. (1974). Conditional logit analysis of qualitative choice behavior. In P. Zarembka (Ed.), *Frontiers in econometrics* (pp. 105–142). New York: Academic.

Papadiochou, O., Triandafyllou, M., & Damigos, D. (2011). *Estimation of the value of water in the Asopos basin using contingent valuation method.* MSc dissertation, National Technical University of Athens, School of Mining and Metallurgical Engineering. (In Greek)

Passali, D. (2009). *Industrial pollution and community-based mobilization in the Asopos River.* MA Dissertation, Department of Geography, King's College London.

Rose, J. M., & Bliemer, M. C. J. (2009). Constructing efficient stated choice experimental designs. *Transport Reviews, 29*, 587–617.

Sándor, Z., & Wedel, M. (2001). Designing conjoint choice experiments using managers' prior beliefs. *Journal of Marketing Research, 38*, 430–444.

Scarpa, R., Notaro, S., Louviere, J., & Raffaelli, R. (2011). Exploring scale effects of best/worst rank ordered choice data to estimate benefits of tourism in Alpine Grazing Commons. *American Journal of Agricultural Economics, 93*(3), 813–828.

Chapter 7
A Value Transfer Approach for the Economic Estimation of Industrial Pollution: Policy Recommendations

P. Koundouri, N. Papandreou, M. Stithou, and O.G. Dávila

Abstract In order to make possible an economic estimation of industrial pollution, which is one of the main polluters in the Asopos River Basin (RB), another method that of Benefit Transfer (BT) was applied and is presented in this chapter. The fact that gathering primary site-specific data is costly and time-consuming has made BT a more and more popular alternative for the valuation of ecosystem goods and services and it offers a considerable potential in the light of the EU Water Framework Directive (WFD) implementation. In a broad sense, BT method uses existing economic value estimates from one location to another similar site in another location. In this context, the objective of this chapter is to present an empirical application of the methodology of transfer value. A number of valuation studies in the European territory that have explored the impact of industry on water degradation are reviewed in order for a suitable 'match' to be made between the Asopos RB and a suitable existing valuation study from which to source economic value information and hence perform the valuation exercise. The chapter closes with conclusions and recommendations for policy design.

Keywords Benefit Transfer • Asopos River Basin • Water Framework Directive • Industrial pollution • Water quality and quantity

P. Koundouri (✉) • M. Stithou • O. Dávila
Athens University of Economics and Business, Athens, Greece
e-mail: pkoundouri@aueb.gr; stithoumavra@yahoo.com; osiel.davila@soas.ac.uk

N. Papandreou
Andreas G. Papandreou Foundation, Athens, Greece
e-mail: tulayev@gmail.com

P. Koundouri and N.A. Papandreou (eds.), *Water Resources Management Sustaining Socio-Economic Welfare*, Global Issues in Water Policy 7,
DOI 10.1007/978-94-007-7636-4_7,

1 Introduction

As it was presented in Chap. 3 the broad Asopos area is the largest industrial region of Greece, supporting 1,300 industrial facilities. It is reminded that in 1969 under a Presidential Decree, the dictatorial regime provided the industries in Athens with incentives in order to transfer their businesses in the Viotia region. The area started at that time to receive several industries, and Asopos River was proclaimed receiver of treated industrial waste. What characterized the location of those industries in the region was the lack of any planning and monitoring system for the control of the area's industries. In 1979, the Prefectures of Attica and Viotia determined the terms and conditions under which the industries would discharge their waste in the river, however, without establishing any monitoring system for the enforcement of those terms. Although today these terms are obsolete and in opposition with more recent laws and regulations, the treaty is still in effect. In addition, more permissions were granted for the operation of industries in the area of the region where the construction was forbidden under the 1969 Presidential Decree. Therefore, hundreds of industrial facilities have been dumping toxic waste in the Asopos River for decades, and as a result, the river and groundwater have been subject to long-term industrial pollution. According to Loizidou's study (2009) there are about 10,500 m^3 of industrial waste, 15 tones of organic waste, which fall on a daily basis in the river and which are not all degradable. Only about 5 tones are degradable while the rest is accumulated and creates serious problems in the wider area.

The most important industrial sector in the area is that of metallurgy and then the sector of food industries, the sector of plastic products and the sector of textile/dyeing/finishing. These sectors are responsible for the main flow of the produced industrial wastewaters. The existing ways of wastewater disposal in the area are the surface disposal, the underground disposal, the recycling of the treated out flow (within production activity), the disposal in a municipal wastewater treatment plant or in an authorized management body and the disposal in the Asopos River or its tributaries. Industrial sector in the area is responsible that waters along the river and at the coastal area are polluted with inorganic and organic load while its contribution to the pollution of groundwater with hexavalent chromium Cr (VI) is considerable. It is regarded that from 2005 to 2009 in the wider area of Asopos there have been imposed fines of €3,424,620 (Technical Chamber of Greece 2009).

The WFD 2000/60/EC (CEC 2000) sets the environmental quality standards at Community level. In particular, the chemical status is required to meet the environmental objectives for surface waters established in Article 4(1) (a). European Union legislation provides for measures against chemical pollution of surface waters in two levels – with Community wide selection of substances of concern and Community wide measures and a requirement that Member States take measures at river basin level against relevant pollutants. There is currently a transitional period until the year 2013 from the "old" framework of Directive 76/464/EEC to the new WFD.

The major part on Community strategy against pollution of surface waters control policy is set out in Article 16 of the WFD which requires the establishment of a list of priority substances and a procedure for the identification of priority substances/priority hazardous substances as well as the adoption of the specific measures against pollution with these substances. The Directive sets the procedure for the setting of chemical quality standards by Member States and expresses monitoring concerns.

The Priority substances Directive (Directive 2008/105/EC) setting environmental quality standards for the priority substances and certain other pollutants is the result of the requirements set in Article 16(8) of the WFD. In addition, the Annex II to this new directive replaces Annex X of the WFD referring to the list of priority substances. Member States shall take actions to meet those quality standards by 2015 as part of chemical status (Article 4 and Annex V point 1.4.3). For this purpose a program of measures (according to Article 11) shall be in place by 2009, and become operational by 2012. In Annex I of this Directive, Environmental Quality Standards (EQS) for priority substances and certain other pollutants (about 33) are provided. Information concerns annual average, maximum allowable concentration while the unit employed is μg/l.

As the primary objective of this chapter is to determine the value of improvements to the quality and quantity of water resources which are facing considerable pressure due to industrial activity, the adopted approach includes an analysis of the impacts of industry on the environment and society beyond economic activities (welfare impacts). After impacts have been identified the focus is on quantifying as many of these as possible, for example the number of people affected or the extent of the area affected. We then move to an economic valuation, adapting values from existing literature to estimate the impacts of intervention in monetary terms. Conclusions and policy recommendations are offered at the end of the chapter where transfer values are compared with the cost of creating a central wastewater treatment plant in the Asopos area.

2 Identification of Impacts of Industrial Pollution

The main objective of this chapter is the valuation of benefits associated with mitigation policies and measures that aim to improve matters in Asopos RB. At a first instance, the impacts will be felt on the environmental goods and services provided by the area, such as amenity, clean water and biodiversity. Generally, these are not traded in markets and consequently no market price is available to reflect their economic value. Values must therefore be derived and, for this reason, environmental resources are increasingly becoming defined in terms of the ecosystem services they provide. Generally, services include provisioning services (products obtained from ecosystems), regulating services (e.g. climate regulation, water regulation), cultural services (e.g. aesthetic values, recreation) and supporting services necessary for the production of all other services. The following table

Table 7.1 Identification of environmental and social impacts of industrial pollution in Asopos catchment

	Impacts	
Asopos RB		
	Environmental:	Biodiversity (fish and invertebrates, birds on estuary)
		Pollution and reduction of groundwater
	Social:	Human health (from polluted agricultural products and groundwater consumption)
		Cost on local economy: increased cost for drinking water for households, increased cost for local agricultural producers, increased cost for food industries, decrease of tourists for local tourist companies
		Recreation (local residents and visitors)
Water bodies within Basin district		
River	Environmental:	Biodiversity (fish and invertebrates)
	Social:	Human health (from polluted agricultural products and water consumption)
Oropos lagoon and coastal zone	Environmental:	Biodiversity (birds on estuary and Oropos lagoon)
	Social:	Recreation (local residents and visitors)
Groundwater	Social:	Human health (from polluted agricultural products and groundwater consumption)

(Table 7.1) presents the estimated impacts of industrial sector on Asopos RB revealing environmental and social impacts.

The baseline scenario is the current situation (*status quo*) in Asopos RB without any intervention, which is the "No Change" scenario. Under this scenario the ecological status of the basin is bad and high levels of pollution are related to serious health concerns. The alternative scenario assumes that mitigation measures (central wastewater treatment in the industrial zone or a build-in innovative technology for industries) to address environmental degradation and human health concerns are taken forward, which is the "Intervention Project" scenario. This is expected to result primarily in social and environmental impacts as well as in financial impacts on industry's production (foregone cost). It should be noted that the emphasis of this chapter is on the first category of impacts.

3 Benefit Transfer Methodology

A more cost-effective approach for the valuation of water quality improvements is expected to come through the application of BT. The fact that gathering primary site-specific data is costly and time-consuming has made BT a more and more popular alternative for the valuation of ecosystem goods and services. The potential of BT has been explored in a number of studies. An example is that of Johnson et al. (2008) who used BT in a stated preference study in England and Wales in order to calculate public WTP for a reduction in risk of illness resulting from swimming in contaminated river waters in Scotland. The study was framed in the context of the EU Bathing Waters standards and WFD. Furthermore, the application of BT in the context of the WFD has been examined and tested in Hanley et al. (2006a, b) by applying Choice Experiment (CE) in two similar rivers and then exploring the possibility of using BT. Results from the two studies are different proving that BT is not a straightforward task to be applied in every case.

BT method uses existing economic value estimates from one location to another similar site in another location. In particular, it concerns an "application of values and other information from a 'study' site where data are collected to a 'policy' site with little or no data" (Rosenberger and Loomis 2000, p. 1097).

Bergland et al. (1995) discussed three main approaches to BT: (i) the transfer of the mean household WTP (ii) the transfer of an adjusted mean household WTP and (iii) the transfer of the demand function. Hence, while the first approach assumes similarity in good's and socioeconomic characteristics between the study and target site, the other two approaches attempt to adjust the mean WTP and re-calculate it respectively, in order to account for differences between the two sites in terms of environmental characteristics and/or socioeconomic characteristics. More particularly, in the case of unadjusted mean value transfer the H_0 is: $WTP_{study\ site\ (s)} = WTP_{policy\ site\ (p)}$. On the contrary, the adjusted value transfer tests the hypothesis: *predicted* $WTP_p(\beta_s, X_p) = WTP_p$, where predicted $WTP_p(\beta_s, X_p)$ is the WTP at the policy site estimated using the parameters of the benefit function of the study site (β_s) and the X values (site attributes, socio-economic characteristics etc.). In the case of benefit function transfer, the value function estimated for the study site is transferred to the policy site and the relevant test concerns the comparison of function parameters between sites: $\beta_s = \beta_p$.

It should be noted as well that meta-analysis can be used to inform the BT processes (Hanley et al. 2006a). When data are pooled across study sites to produce a BT model for predicting policy site values, the test is: $\beta_{s+p} = \beta_s$ and $\beta_{s+p} = \beta_p$ where, β_{s+p} are the parameters of the pooled regression models.

Generally, the benefit function option seems to be preferred as among others it accounts for differences in site characteristics and human populations between sites. However, function transfers are "limited by quality and availability of primary research, limited consensus on performance and validity of types of function transfers and lack of consensus on how to generate functions"[1] (Rosenberger and Johnston 2009). It should be also noted that in terms of

[1] R.S. Rosenberger's presentation on "Methods, Trends, and Controversies in Contemporary Benefit Transfer" joint paper with Robert J. Johnston. COST E45 EUROFOREX Training Course May 13–17 2009, Benefit Transfer – Introduction and Methods, Norway.

Transfer Error (TE) [2] function transfer does not seem to perform better than unit value transfer.

4 Findings of Literature Review and Selection of the Economic Value Evidence

It is a fact that we would expect values for a particular good to differ between different locations, for a variety of good reasons relating to context and the particular characteristics of an area. However, in order to minimize concerns, it is important to select the most appropriate value estimate from the most appropriate study site or good. Current best practice in terms of the application of value transfer to a specific project suggests that a number of key points should be taken into consideration when deciding which piece(s) of economic literature can provide the 'best value estimate' for a specific ecosystem service or habitat change.

We have conducted a thorough and detailed review of the economic valuation literature relating to benefit transfer and to industrial impacts in particular. The transferability of each study to the current situation has been assessed, with a focus on the good in question, the environmental change considered, the population and location. The number of valuation studies worldwide related to river related improvements is quite extensive. In order to narrow down our search for estimates so as to not to comprise reliability and validity of results the following criteria were followed. Considered studies should have in common: (i) the stressor (ii) the baseline scenario and change in the provision of the good/service (iii) the affected population both in terms of its characteristics and geographical extent. Therefore, the study site should be a severely polluted mainly rural river affected considerably by industrial activities. In addition, the two catchments should demonstrate similar socio-economic characteristics which however typically vary between different countries, most notably when comparing developing and more developed countries.

We have considered studies undertaken in the European territory and Mediterranean whenever possible. It has to be noted as well that in general only a limited number of valuation studies have been undertaken in Greece and they generally relate to a very different kind of good from that of interest here. In addition, an effort was made in order to find studies that are motivated by the WFD and therefore share the same policy framework. Finally, scale is another issue of concern as a number of studies consider regional scale for example "rivers and lakes of East Anglia" (Bateman et al. 2006) or even national level (Baker et al. 2007).

[2] TE is defined as the percent difference between the transferred-predicted (WTPT) and policy site-observed primary estimate (WTPP):

$$TE = \frac{|WTP_T - WTP_P|}{WTP_P}$$

However, it is not underestimated the fact that studies from other countries generally involve significant differences between the ecology of resources in the study sites, the affected populations and their socio-economic characteristics and those of the population within the Asopos RB. Often these studies are therefore not a good match with the current project. A number of studies have sought to estimate the value of social and environmental impacts arising from industrial activity. Several of the studies relate to a particular resource, the results of which are driven by specific attributes of the local area making them less desirable for inclusion within a value transfer. The findings of our literature review include two main categories and focus mainly on river freshwater degradation. However, a less extensive literature review encompasses also impacts on groundwater and wetland. The first category is related to *environmental impacts* caused by industrial pollution. The literature review includes:

(a) Studies from European territory related to industrial pollution
(b) Studies motivated by the WFD regardless of source of pollution
(c) Studies having water element (in general) in Greece

The second category is related to *human health impacts* caused by industrial pollution. In addition, as previously mentioned literature review includes groundwater pollution from industrial and wetland pollution in order to consider Oropos lagoon degradation. An important instrument of our literature review was the EVRI[3] database. Several of the studies although relate to river quality their results are driven from various attributes of the local area making them less desirable for inclusion within a value transfer. Furthermore, there is a considerable group of studies that relate recreation to river improvements that have not been included as not such an interest is expressed in Asopos at the moment.

Table 7.2 presents valuation studies in the European territory where industry contributed considerably to water degradation.

Hanley et al. (2006a), using CE analyzed improvements to the ecology of the River Wear, in Durham, England and the River Clyde, in Central Scotland. These rivers were chosen as broadly representative of the kind of water bodies in the UK where moderate improvements in water quality are likely to be needed in order to meet good ecological status. The lower sections of the River Wear were heavily polluted by industry and mining, but have now recovered and support a migratory fishery. Existing problems include litter, algal growth and acidity problems due to mine drainage. Problems also exist with loss of bankside vegetation, increased erosion, and a decline in habitat and associated fish and wildlife populations. The river plays an important role in recreation and tourism. The River Clyde has recreational and tourist attractions, and encompasses areas of great beauty Falls of Clyde, but also has some of the most problematic stretches in terms of water quality. Most of this section was graded 'B' using the Scottish river classification system, which is equivalent to the 'C' grade for the Wear under the General Quality

[3] http://www.evri.ca

Table 7.2 Overview of studies considered for benefit transfer

Authors	Country	Water body	Method	Mean WTP values (per household/year)
Hanley et al. (2006a)	Scotland	Catchment river	Choice experiment	£20.17 for the improvement of the ecology of the river from 'fair' to 'good'
Hanley et al. (2006b)	Scotland	Catchment river	Choice experiment	£8.97 for the improvement of the ecology of the river from 'fair' to 'good' and £24.03 for the improvement of the ecology of the river from 'fair' to 'very good'
Martin-Ortega et al. (2009)	Spain	Catchment river	Choice experiment	€81.2 for the improvement of the ecology of the river from 'bad' to 'very good' and 61.3€ for the improvement of the ecology of the river from 'bad' to 'good'
Kataria et al. (2009)	Denmark	Catchment river	Choice experiment	491 DKK for the improvement of the ecology of the river from 'bad' to 'good' and 547DKK to 'very good'
Bateman et al. (2006)	England	Catchment river	Contingent valuation	£15.24 for the improvement of the ecology of the river from 'bad' to 'good' and £22.89 to 'very good'
Birol et al. (2008)	Cyprus	Wetland	Contingent valuation	18.25 cyp for the scenario of the maximum improvement of the ecology of the wetland
Birol et al. (2006)	Greece	Wetland	Choice experiment	14.45 € for the scenario of the maximum improvement of the ecology of the wetland
Carlsson et al. (2003)	Sweden	Wetland	Choice experiment	493.76 SEK for the improvement of the ecology of the wetland from 'bad' to 'good 'and 719.75 SEK from 'bad' to 'very good'
Birol et al. (2010)	Cyprus	Groundwater	Choice experiment	0.014 CYP/per m^3 for the improvement of water quality to the maximum
Rinaudo (2003)	France	Groundwater	Contingent valuation	77€ for the restoration of the very bad ecological condition
Bergstrom et al. (2004)	USA	Groundwater	Contingent valuation	$47.81 for decontamination of nitrate loads

Assessment classification system (i.e. fair quality, but in need of improvement to reach 'good ecological status').

In their paper, Hanley et al. (2006b) analyzed the case of two small catchments located in eastern Scotland: the Motray and the Brothock. This area has difficulties in meeting Good Ecological Status because of the presence of high nutrient levels

(N and P) and low summer river flows. The use of fertilizer and manure applications by farmers is the cause of the high nutrient levels. They use CE in order to estimate willingness to pay (WTP) for improvements in the ecological status of the catchments (through stricter controls on irrigation and on diffuse-source pollution).

Martin-Ortega et al. (2009) used a choice experiment based on maps in order to elicit welfare measures for water quality improvements across sub-basins in the Guadalquivir River Basin in Spain. The Guadalquivir River Basin (GRB) is the longest river in the south of Spain. Water quality is a significant problem throughout the river basin. The main sources of pollution include urban and industrial wastewater discharge, erosion, nutrient and pesticide runoff from agricultural land. Concentration levels of Nitrogen, Phosphorus, heavy metals and organic pollutants in surface and ground waters are expected to increase with about 30 % in the near future.

Kataria et al. (2009), estimate the WTP for water quality improvements in the Odense River in Denmark using a CE study. The present quality of Odense River is affected by human activities and is classified as moderate. They found that the majority of their respondents find the improvement in the scenario described to them to be unlikely or rather unlikely. Thus, it appears to be a mistrust of the environmental improvements described in their survey, especially when it comes to achievement of the best water quality.

In their paper, Bateman et al. (2006) conducted a valuation of improvements to the water quality of the River Tame. This is an urban river that flows through the city of Birmingham in the UK and is classified as having very poor water quality by the UK Environment Agency. Fish stocks are virtually non-existent and other fauna and flora are severely limited. Direct human use is seriously limited with the river classified as being unsuitable even for boating. Nevertheless, the river has an ecological and recreational potential and passes through residential areas, playing fields and a country park. In the paper, applications of contingent valuation (CV) and contingent ranking (CR) methods were used.

Birol et al. (2006), used a choice experiment to estimate the values of changes in several ecological, social and economic functions that Cheimaditida wetland provides. This wetland includes Lake Cheimaditida, one of the few remaining freshwater lakes in Greece. It provides several of the important ecological functions. The value of the economic benefits generated by sustainable management of the wetland was estimated using data from 407 CE surveys that were administered in 10 cities and towns in Greece. Within the wetland the main economic activities include agriculture, forestry and fishing. Agriculture is a vital activity where alpha-alpha and maize are the main cash crops whose production is water and fertilizer intensive. Water opportunities from the lake for irrigation in agriculture, and pollution due to run-off from agricultural production, have adverse effects on water quantity and quality. These in turn affect the level of biodiversity that the wetland is able to support.

Birol et al. (2008), assessed an aquifer management plan to replenish a depleting aquifer with treated wastewater in Cyprus (a water-scarce region of the world). They conducted two distinct CEs on randomly selected members of two key stakeholder groups, allowing them to estimate the use and non-use economic benefits that may arise as a result of the proposed aquifer management plan. This aquifer faces water quality and quantity problems. Since the construction of the Kouris River dam, the aquifer's water inflow has decreased significantly, resulting in a lower water table. This has led to the intrusion of saltwater as the aquifer attempts to maintain its hydrological balance. Water quality in the aquifer has further deteriorated due to the intensive use of fertilizers and pesticides in the area's agricultural production. The quantity of water in the aquifer has been reduced due to uncontrolled and excessive pumping.

Birol et al. (2010) present their progress on the research of the Akrotiri aquifer. In this paper, they provide a methodology for assessing the viability of an environmental management plan to replenish this depleting aquifer with treated wastewater. The plan has long-run economic and ecological impacts. They conducted two distinct choice experiments in order to capture the different components of economic value of two stakeholder groups: local farmers and public located in the nearby city of Limassol. Their results reveal that the farmers and residents are not opposed to an aquifer management plan that proposes to replenish the aquifer with treated wastewater.

Birol et al. (2009) used a choice experiment to estimate the value of management options for the Bobrek wetland in Poland. It was implemented in the city of Sosnowiec, located in the Bobrek catchment in the Upper Silesia region of Poland. The region is an important industrial center located within the Upper Silesian Coal Basin. Five rivers run through the wider area, including the Biala, Brynica, Jaworznik, Wielonka and Rawa, making the region susceptible to flooding episodes. Extensive mining activities generated solid waste dumping in the form of spoil heaps which resulted in the degradation of the aquatic and terrestrial ecosystems. Waste dumping has also resulted in extensive pollution from heavy metals and other pollutants in this area.

Bergstrom et al. (2004), developed a conceptual model in order to analyze how the different payment vehicles of a special tax and a tax reallocation affects the WTP for ground water quality protection in Georgia and Maine, US. Their results show that WTP with a tax reallocation is higher than WTP with a special tax for ground water quality protection.

Carlsson et al. (2003), estimated individuals' marginal willingness-to-pay (WTP) for different attributes of a wetland in Staffanstorp, southern Sweden. In this area there has been a public discussion about the location, design and construction of a wetland. The respondents were the local population living in Staffanstorp. They found a negative mean WTP for the three attributes "Meadow land", "Fenced waterline" and "Crayfish". They conclude that an inclusion of these attributes will decrease social welfare.

Finally, Rinaudo (2003) analyzed the case of a highly polluted area of the upper Rhine valley alluvial aquifer. The groundwater in this aquifer has been affected by different kinds of pollution. High concentrations of Nitrate, chlorinated hydrocarbons and pesticide pollution problems have been reported. The chloride pollution is provoked by the potash mining industry. For the drinking water sector, the estimated value of the economic damage is 17.5 M€ (60 % of the total cost) and 5.5 M€ for the rest. Using the results of an existing contingent valuation study the economic value of the pollution of the aquifer was estimated at 6.6 million €.

5 Valuation Exercise

As it was stated in the previous section, the main criteria for the selection of the studies were the (i) stressors (ii) the baseline scenario and (iii) the population. In the case of the valuation of river water quality, the paper by Martin-Ortega et al. (2009) was selected as the most appropriate study to be used in order to conduct a BT valuation exercise. The Guadalquivir River Basin has a Mediterranean climate. The study was motivated by the European WFD and therefore shares the same policy framework as the Asopos RB. In relation to the stressors, the main sources of pollution include urban and industrial wastewater discharge as in the Asopos RB.

Hanley et al. (2006a) is not appropriate since the rivers in their study are polluted not only by industry but also mining (the existing problems include litter, algal growth and acidity problems due to mine drainage), and have now recovered and support a migratory fishery. The mining feature is not shared by the Asopos RB. In the same way, Hanley et al. (2006b) is also not an appropriate study because the problems in their case study areas are high nutrient levels and low summer river flows. Problematic nutrient levels are mainly due to fertilizer and manure applications by farmers. Thus, the stressor's nature is different. The paper by Kataria et al. (2009) was rejected because it does not convey precise information about the stressors. On the other hand, the water quality in this river is classified as moderate (the river is suitable for boating with limited possibilities for swimming and angling. Pollution sensitive fish species are present but the presence of birds, plants and other fish species is limited). The baseline scenario is different: it values environmental improvements for the hydropower regulated rivers in Sweden. Bateman's (2006) study is not appropriate as the environmental good is not the same-urban river and the size of the catchment is very big and the sites where the goods are found are not the same as in the Asopos RB.

In the case of the valuation of wetland quality, Birol et al. (2006) was selected as the most appropriate study. They estimated values of changes in ecological, social and economic functions in Cheimaditida wetland. This wetland includes Lake Cheimaditida, one of the few remaining freshwater lakes in Greece and it provides important ecological functions.

Although Birol et al. (2008) could be considered appropriate because the baseline is quite similar they assessed an aquifer management plan to replenish a depleting aquifer with treated wastewater in Cyprus (a water-scarce region). Thus, Birol et al. (2006) is preferred above Birol et al. (2008). Finally, Carlsson (2003) was also not considered appropriate because the study focus is on the construction of wetlands.

In relation to groundwater valuation, the paper by Bergstrom et al. (2004) was not selected because the study area is mostly urban and not comparable with the Asopos RB. In the same fashion, the paper by Rinaudo (2003) was not selected because the source of pollution originates from potash mining waste dumps. Thus,

Birol et al. (2010) was selected as the most appropriate study. The study was motivated by the European WFD and the valuation was about an aquifer management plan to replenish a depleting aquifer in a water-scarce region.

In this valuation exercise the Unit Value Transfer method is used. For the wetland case a simple unit value transfer is used. For the river and groundwater cases, a unit value transfer with adjustment for income differences is used (because the simple unit value transfer approach should not be used for transfer between countries with different income levels and costs of living). The value estimate is therefore adjusted from the time of data collection to current currency using the Consumer Price Index (CPI) for the policy site country (Greece). Since values are transferred from a study site outside the policy site country, first the values are converted to reflect the purchasing power in the year of data-collection, using PPP (Purchase Power Parity) corrected exchange rates in the year of data collection, and then the local CPI to update to current-currency values is used. Taking this into consideration, the benefit estimate (expressed as mean willingness-to-pay (WTP)/household/year) from the study site to the policy site (Asopos RB) is then transferred.

Following Navrud and Ready (2007), the adjusted WTP estimate B_p' at the policy site was calculated using the following equation: $\mathrm{WTP}B_p' = \mathrm{WTP_s}\ (\mathrm{Y_p}/\mathrm{Y_s})^{\beta}$ where WTP_s is the original WTP estimate from the study site, Y_s and Y_p are the income levels at the study and policy site, and β is the income elasticity of demand for the environmental good in question. Income elasticity of WTP β for different environmental goods are typically smaller than 1, and often in the 0.4–0.7 range. Navrud reports a multi-country CV study of Value of a Life Year (VOLY) that found the income elasticity to be about 0.2 and 0.5 for the EU-15 and the New Member Countries, respectively. In this valuation exercise an income elasticity of 0.5 is used. Gross Domestic Product (GDP) per capita figures have been used as proxies for income.

GDP per capita based on purchasing power parity (PPP) figures were obtained from the World Bank, International Comparison Program database. The GDP per capita (PPP) for Spain in 2006 (the year of data collection) was 30,333. The GDP per capita (PPP) for Greece in 2006 was 26,733. The GDP per capita (PPP) for Cyprus in 2008 (the year of data collection) was 31,816. The GDP per capita (PPP) for Greece in 2008 was 29,604. Table 7.3 presents the mean values for benefits of

Table 7.3 Mean WTP values (per household/year)

Martin-Ortega et al. (2009)		Birol et al. (2006)	Birol et al. (2010)
Improvement of the ecology of the river from 'bad' to 'very good'	Improvement of the ecology of the river from 'bad' to 'good'	Maximum improvement of the ecology of the wetland	Improvement of groundwater quality to the maximum
€81.20	€61.30	€14.45	CYP 0.014 per m^3
Unit transfer with income adjustments[a]			
€116.94	€ 88.28	€14.45	€0.021 per m^3

[a]GDP per capita based on purchasing power parity (PPP) figures and GDP deflator for Greece (base year 2005=100) were obtained from the World Bank, International Comparison Program database. PPP (Purchase Power Parity) corrected exchange rates were obtained from OECD.STAT. Figures are expressed in 2005 €

improving the water status for groundwater, wetland and river bodies. The value transfer estimates for improvement of the ecology of the river from 'bad' to 'very good' is €116.94. The value transfer estimate for maximum improvement of the ecology of the wetland is €14.45. Finally, the value transfer estimate for improvement of groundwater quality to the maximum is €0.021 per m^3.

6 Conclusions

In this chapter in order to assess the welfare impacts associated with improvements in water bodies of Asopos basin we are using existing valuation literature and our expertise in transferring values from relevant and comparable projects. During the selection of the economic value evidence a number of conditions were established. Only studies with a common environmental stressor (industrial pollution) were considered. They should value similar impacts, should share similar baseline scenario and good and the location should be as close to policy site as possible. Therefore, only studies undertaken in the European territory and Mediterranean were used. This was one of the main challenges since a limited number of valuation studies have been undertaken in Greece and they generally relate to a very different kind of good. An effort was made to find studies that are motivated by the WFD and therefore share the same policy framework. The findings of our literature review include two main categories and focus mainly on river freshwater degradation and also impacts on groundwater and wetland. The first category is related to environmental impacts caused by industrial pollution. The literature review includes studies from European territory related to industrial pollution, studies motivated by the WFD regardless of source of pollution and studies having water element (in general) in Greece. The literature review includes groundwater pollution from industrial and wetland pollution in order to consider Oropos lagoon degradation. Several of the studies although relate to river quality their results are driven from various attributes of the local area making them less desirable for inclusion within a value transfer.

The Unit Value Transfer method was used in the valuation section. For the river and groundwater cases, a unit value transfer with adjustment for income differences was used. The value estimate was adjusted from the time of data collection to current currency using the Consumer Price Index (CPI) for the policy site country (Greece). A simple unit value transfer was used for the wetland case. The obtained value transfer estimate for improvement of the ecology of the river from 'bad' to 'very good' is €116.94. The value transfer estimate for maximum improvement of the ecology of the wetland is €14.45. Finally, the value transfer estimate for improvement of groundwater quality to the maximum is €0.021 per m^3.

However, it should be noted that adding up estimates from separate studies on the value of various water bodies that may impact on the same ecosystem might result in some double counting of benefits. Furthermore, it should be also considered that the fact that substitution effects and budget constraints are often incompletely accounted for, may lead to over-valuation. Nevertheless, trying to aggregate values to the households of Asopos in order for example to estimate the social benefit from moving the river's ecology from the *status quo* (baseline scenario) to "*Good Ecological Status*" (GES) as WFD dictates, reveals a value of about €2,690,000 per year considering that within Asopos RB there are 23,000 households which are permanently connected with the public system of water supply. The respective value for achieving a maximum improvement in the ecology of the wetland is about €332,000 per year. Contrasting the above estimate for river improvements to results of Chap. 6 regarding the moderate environmental improvements in all water bodies valued at €882,200 per year, we can see that the BT value is overestimated.

As far as costs of interventions are concerned, a measure suggested by Loizidou (2009) in order to alleviate the problem in the area was the construction of a Central Wastewater Processing Unit for the disposal of industrial and urban wastewaters. The total cost for the construction was estimated at €32,430,533 while the running cost was ≈ €0.611 per m^3 of waste. The annual running cost for an average daily provision of 16,762 m^3 was 0.611*365*16,762 ≈ €3,738,177. In addition, the average depreciation cost (€/m^3) of industrial waste (considering 40 years) was estimated at €0.137 per m^3. Dimaras et al. (2010) study also reports that only the construction of a new pipeline connection to provide Oropos area with clean water was estimated at €9,400,000.

Overall, the reported estimates in this chapter provide additional evidence of the considerable welfare loss associated with the environmental degradation of Asopos RB as a result of unplanned industrial development. The estimated values provide a considerable contribution against the cost of necessary mitigation measures and show as well that the fines paid by the industry underestimate the cost imposed on the society (ecological, human health, forgone earnings) with fees being totally non-reflective of the pollution cost.

Therefore, our results can be used in the design of public policy in the area. The estimated values that people would place on improvements to the river's ecology (as envisaged under the WFD) suggest that social welfare can be increased by establishing a water/pollution management plan to control pollution in the Asopos RB.

Apart from close monitoring, regular inspections and audits, within this plan it is important that regulatory stakeholders adopt modes of operation in line with the opportunities presented within the full cost recovery options using polluter and user pays principle. Making the polluters accountable for the ecological damage they are causing to themselves and to their future generations and calculating revenue amounts for water services keeping in mind ecological costs is expected to lead to a more effective management. The polluter pays principle relies on incentive-based water pricing that will lead to behavioral changes for example, installing green technologies and on correct economic cost allocation. Hence, imposing "Green" taxes/penalties/subsidies for polluting industries, depending on their pollutant loads, quantities and willingness to adopt environmental friendly practices, can enable local government collecting and generating new pollution revenues to be reinvested in the pollution management plan.

References

Baker, B., Metcalfe, P., Butler, S., Gueron, Y., Sheldon, R., & East, J. (2007). *The benefits of Water Framework Directive programmes of measures in England and Wales*. A Final Report to DEFRA re CRP Project 4b/c. NERA Economic Consulting and Accent.

Bateman, I. J., Cole, M. A., Georgiou, S., & Hadley, D. J. (2006). Comparing contingent valuation and contingent ranking: A case study considering the benefits of urban river water quality improvements. *Journal of Environmental Management, 79*(3), 221–231.

Bergland, O., Magnussen, K., & Navrud, S. (1995). *Benefit transfer: testing for accuracy and reliability* (Discussion Paper#D-03/1995), Department of Economics and Social Sciences, The Agricultural University of Norway, Norway.

Bergstrom, J. C., Boyle, K. J., & Yabe, M. (2004). Trading taxes vs. paying taxes to value and finance public environmental goods. *Environmental and Resource Economics, 28*(4), 533–549.

Birol, E., Karousakis, K., & Koundouri, P. (2006). Using a choice experiment to account for preference heterogeneity in wetland attributes: The case of Cheimaditida wetland in Greece. *Ecological Economics, 60*(1), 145–156.

Birol, E., Koundouri, P., & Kountouris, Y. (2008). Integrating wetland management into sustainable water resources allocation: the case of Akrotiri wetland in Cyprus. *Journal of Environmental Planning and Management, 51*(1), 37–53.

Birol, E., Hanley, N., Koundouri, P., & Kountouris, Y. (2009). Optimal management of wetlands: Quantifying trade-offs between flood risks, recreation, and biodiversity conservation. *Water Resources Research, 45*(11).

Birol, E., Koundouri, P., & Kountouris, Y. (2010). Assessing the economic viability of alternative water resources in water-scarce regions: Combining economic valuation, cost-benefit analysis and discounting. *Ecological Economics, 69*(4), 839–847.

Carlsson, F., Frykblom, P., & Liljenstolpe, C. (2003). Valuing wetland attributes: an application of choice experiments. *Ecological Economics, 47*, 95–103.

Commission of the European Communities (CEC). (2000). *Directive of the European Parliament and of the council establishing a framework for community action in the field of water policy*. 1997/0067 (COD), C5-0347/00.

Dimaras, A., Mastrogiannis, F., & Damigos,D. (2010). *Estimation of the cost of pollution of Asopos River*. MSc dissertation, National Technical University of Athens, School of Mining and Metallurgical Engineering. (In Greek)

Directive 2008/105/EC of the European Parliament and of the Council of 16 December 2008 on environmental quality standards in the field of water policy, amending and subsequently repealing Council Directives 82/176/EEC, 83/513/EEC, 84/156/EEC, 84/491/EEC, 86/280/EEC and amending Directive 2000/60/EC of the European Parliament and of the Council.

Hanley, N., Wright, R. E., & Alvarez-Farizo, B. (2006a). Estimating the economic value of improvements in river ecology using choice experiments: an application to the water framework directive. *Journal of Environmental Management, 78*, 183–193.

Hanley, N., Colombo, S., Tinch, D., Black, A., & Aftab, A. (2006b). Estimating the benefits of water quality improvements under the Water Framework Directive: Are benefits transferable? *European Review of Agricultural Economics, 33*(3), 391–413.

Johnson, E. K., Moran, D., & Vinten, A. J. A. (2008). A framework for valuing the health benefits of improved bathing water quality in the River Irvine catchment. *Journal of environmental management, 87*(4), 633–638.

Kataria, M., Hasler, B., Christensen, T., Martinsen, L., Nissen, C., Levin, G., Dubgaard, A., Ladenburg, J., Bateman, I., & Hime, S. (2009). *Scenario Realism and Welfare Estimates in Choice Experiments – Evidence from a study on implementation of the European Water Framework Directive in Denmark*. Paper presented at the EAERE 17th Annual Conference, Session: Stated Preferences 5, 24–27 June 2009, Amsterdam, The Netherlands.

Loizidou, M., (2009) *Environmental impact assessment for a central processing unit for the industrial wastewater of Asopos area and the urban wastewater of the municipality of Avlonas*. National Technical University of Athens, School of Chemical Engineering. (In Greek)

Martin-Ortega, J., Brouwer, R., & Berbel, J. (2009, August) *Economic analysis of spatial preferences heterogeneity of water quality*. Contributed Paper prepared for presentation at the International Association of Agricultural Economists Conference. Beijing, China, pp. 16–22.

Navrud, S., & Ready, R. (Eds.). (2007). *Environmental value transfer: Issues and methods*. Dordrecht: Springer.

Rinaudo, D. J., (2003). *Economic assessment of groundwater protection, Groundwater restoration in the potash mining fields of Alsace, France* (Case study report No. 1. BRGM/RC-52324-FR), Orléans: BRGM.

Rosenberger, R. S., & Johnston, R. J. (2009). Methods, trends, and controversies in contemporary benefit transfer. COST E45 EUROFOREX Training Course, *Benefit Transfer – Introduction and Methods*, May 13–17 2009, Norway.

Rosenberger, R. S., & Loomis, J. B. (2000). Using meta-analysis for benefit transfer: in-sample convergent validity tests of an outdoor recreation database. *Water Resources Research, 36*, 1097–1107.

Technical Chamber of Greece. (2009) *The problem of Asopos River-Suggestions for facing it*. Report, July 2009. (In Greek)

Chapter 8
A Laboratory Experiment for the Estimation of Health Risks: Policy Recommendations

A. Drichoutis, P. Koundouri, and K. Remoundou

Abstract Environmental health hazards in Asopos River Basin (RB) rise concerns on health risk not only for the residents of the catchment but also for the consumers of the area's products across the country. As a result, the focus of this chapter is on the estimation of theses health risks. In order to assess the social cost from consuming products produced in an area where water resources are not in good condition a lab experimental auction has been conducted. According to the details of the experiment presented at the beginning of the chapter all participants were asked to bid to exchange a product from region A (the polluted one) with a similar product from a region in a good ecological status (in the terminology of the European Water Framework Directive (WFD)), region B. After the presentation of the employed methodology econometric analysis was conducted. The output of this analysis is an average Willingness-to-Pay (WTP) for the specific sample of consumers. WTP represents the maximum amount of money a person would be willing to pay in order to receive a good or avoid an undesired impact. The chapter concludes with policy recommendations and suggests that unless an epidemiology study confirms the health hazards from consuming agricultural products cultivated in the area of Asopos (area similar to the characteristics of region A), the estimated amount is a net cost suffered by the local farmers and compensation measures have to be adopted.

A. Drichoutis (✉)
University of Ioannina, Ioannina, Greece
e-mail: adrihout@gmail.com

P. Koundouri
Athens University of Economics and Business, Athens, Greece
e-mail: pkoundouri@aueb.gr

K. Remoundou
Aberystwyth University, Aberystwyth, UK
e-mail: kyriakiremoundou@yahoo.gr

P. Koundouri and N.A. Papandreou (eds.), *Water Resources Management Sustaining Socio-Economic Welfare*, Global Issues in Water Policy 7,
DOI 10.1007/978-94-007-7636-4_8, © Springer Science+Business Media Dordrecht 2014

Keywords Environmental health hazards • Asopos River Basin • Estimation of health risks • Experimental auction • Willingness-to-Pay

1 Introduction

As presented in detail in Chap. 1 of the book, the European Water Framework Directive (WFD) (2000/60/EC) (CEC 2000) is one of the first European Directives to recognize explicitly a role for economics in reaching environmental and ecological objectives. More specifically, the Directive calls for the application of economic principles, economic methods and instruments for achieving good water status for all EU waters in the most effective manner. The most important economic concept that the Directive introduces, is that of water resources management based on the recovery of the total economic cost of water services including the financial cost of water provision as well as the environmental and resource costs. The environmental cost reflects the welfare losses associated with water quality deterioration and the subsequent decrease in water body's capacity to sustain goods and services that are of value to people. However, since many of the services associated with water resources are not marketed, the estimation of the environmental cost is not straightforward. Health risks are an important component of the environmental cost associated with degradation in water quality. Epidemiological studies have found significant morbidity being associated with consumption of products being cultivated with contaminated water. Yet, health risks are often neglected in policy making due to the inherent difficulties in monetizing them.

In this chapter we employ an experimental auction in the laboratory to elicit participants WTP to hedge against health risks associated with consumption of agricultural products cultivated in the Asopos RB. Experimental auctions offer an alternative to stated preference techniques, implemented in Chap. 6, in eliciting preferences and values. Moreover, since subjects are offered real incentives, values are potentially free of hypothetical biases that welfare measures under hypothetical surveys are often criticized for. Health risks due to the use of contaminated water in agriculture are a significant concern in the Asopos RB. Although epidemiological studies relating the degree of pollution with health risks are not available in the area, toxicological surveys have shown that heavy metals concentrations in the groundwater, mainly hexavalent chromium – Cr(VI), are above the safety levels. This presents a potential risk for the health of the people consuming products from the area and subsequently entails a financial cost for the farmers due to difficulties in marketing their products. Therefore results from this experiment aim to feedback in the design of appropriate economic measures to internalize this externality according to the polluter pays principle that the Directive introduces.

2 Review of the Literature

Reflecting greater awareness for environmental and health issues in recent years, a vast literature eliciting valuations for food safety has emerged. The majority of studies use stated preference techniques to elicit people's preferences and WTP premiums for food products that are of certified quality. Genetically modified products and pesticides residues are among the food hazards that have been more extensively studied. The literature in general finds that people are WTP significant amounts to protect themselves from health hazards associated with food consumption. Another common finding is that consumers tend to overestimate low probabilities of risks associated with food safety due to negative publicity and incomplete information (Aldrich 1994). Morkbak et al. (2008) present an overview of the existing studies on the valuation of food safety and quality in meat. In a meta-analysis of studies eliciting preferences toward pesticides-related health risks, Florax et al. (2005) find that people hold significant values for pesticide risk mitigation which are responsive to the proposed risk reduction. There is also a large literature on consumer preferences for genetically modified food. Costa-front et al. (2008) and Lusk et al. (2005) summarize the existing evidence and conclude that in general a high premium for non-GM food is reported in the literature.

Groundwater contamination has not extensively been studied in relation to food safety. Among the exceptions, Stenger (2000) elicit WTP in an experimental setting for protection against heavy metal (cadmium) contamination. Authors stress the importance to account for subjective risk perceptions in the valuation process. In a similar study, Rozan et al. (2004) employ an experimental auction to assess consumers WTP to avoid health risks from heavy metal content in food. Authors find significant premiums that are sensitive to the provided information about food safety.

Drawing on this literature, we examine in this chapter whether consumers are WTP a premium to avoid potential health risks related to heavy metal contamination of the agricultural products. To this end, we designed an experiment where participants had the opportunity to exchange an agricultural endowment from a river basin not in good ecological status according to the terminology of the WFD with a similar endowment from a different region. The experimental setting involved participants bidding in a fourth price Vickrey auction. We are further interested in investigating whether valuation changes when information on the higher health risk to which children are exposed, given their longer time span, is provided. We thus follow a split sample approach with half of our sample being exposed to the additional information. Furthermore we investigate how risk perception influences individual valuations of exposure risk reductions.

3 Experimental Design

3.1 Subjects

We recruited a sample of consumers from the population of Athens. Recruitment was undertaken by a professional research company. Subjects were told that they would participate in an experiment for a compensation of 20 euros for their participation with the understanding that the final amount may vary given decisions during the experiment. The nature of the experiment was not revealed to avoid self-selection. The experiments took place at the laboratory of the Agricultural University of Athens, lasted for about one hour and one of the authors served as the experimenter in all sessions. We conducted four sessions with a total of 61 consumers. Table 8.1 describes the experimental design.

3.2 Procedures and Agricultural Products

For each session a training phase aiming to familiarize respondents with the auction procedures preceded the real auction. Training involved respondents bidding in three hypothetical auction rounds for a bag of potato chips and in three real auction rounds for a chocolate bar. In all sessions a 4th price Vickrey auction was employed. Vickrey auctions are known to be demand revealing in the sense that bidding the true WTP is a dominant strategy (Lusk and Shogren 2007). In the 4th price auction respondents first submit their bids, bids are then ranked and the fourth higher bid is identified. All respondents having submitted bids higher than the 4th highest bid win the auction and buy the product in the fourth highest price. In our auction subjects are first endowed with one kilo of potatoes from what we named river basin A which is in bad ecological status, bearing the same characteristics regarding heavy metals contamination as the Asopos basin. They then bid to upgrade to one kilo of potatoes from what we named river basin B. Basin B was claimed to be certified for being in a *good ecological status* and therefore consumption of products from this area does not involve any health risks. We did not reveal the region of origin of

Table 8.1 Experimental design

Treatments	Subject pool (N)	Information on health risk	Warm glow isolated[a]
1	15	General population	Yes
2	16	General population	No
3	14	General population and children in specific	Yes
4	16	General population and children in specific	No

[a]In two of the sessions, subjects were additionally informed that revenues from higher bidders would be donated in a charity of their choice on their behalf. The design of these sessions aimed to isolate warm glow incentives behind donations. Results from this analysis are reported elsewhere

Table 8.2 Socio-demographic information

Variable	Variable description	Mean (SD)
Bid	Bid to exchange product	0.604 (0.589)
Charity	Dummy, 1=Subject participated in the charitable auction	0.458 (0.502)
HealthRisk	Dummy, 1=Subject received additional health risk information regarding children	0.492 (0.504)
Age	Subject's age	41.508 (9.839)
Gender	Dummy, 1=male	0.305 (0.464)
Income	Dummy, 1=Subject's household economic position is above average	0.475 (0.504)
Kids	Dummy, 1=Subject has kids under 18 years old	0.339 (0.477)
Educ	Dummy, 1=Subject has a university diploma (for the consumers subject pool)	0.610 (0.492)
$PotatoConsumption_1$	Dummy, 1=Subject consumes potatoes 1–2 times/month or less	0.153 (0.363)
$PotatoConsumption_2$	Dummy, 1=Subject consumes potatoes 1 time/week	0.186 (0.393)
$PotatoConsumption_3$	Dummy, 1=Subject consumes potatoes 2–3 times/week	0.441 (0.501)
$PotatoConsumption_4$	Dummy, 1=Subject consumes potatoes 4–5 times/week or more often	0.220 (0.418)

potatoes to avoid any affiliation effects that could potentially bias our estimates. A leaflet was distributed describing the environmental conditions in the areas A and B. The exact text is provided in the Appendix. Subjects participated in five auction rounds and were told that at the end one round would be randomly chosen as binding. In a final stage respondents were asked on their socioeconomics characteristics (age, household size, gender, education, income and whether they have underage children or not) as well as their perceptions of health risks from consumption of products from river basin A and B. Table 8.2 summarizes the socioeconomic profile of the participants.

4 Results

4.1 Risk Perceptions

A large literature in sociology suggests that risk perceptions influence consumer behavior in many different health and food safety contexts and stresses the need to account for perceptions in risk assessment and risk communication (Slovic 1987; Frewer et al. 1998). Our risk perception variables come from replies in the following statements asking people to declare their degree of agreement in a seven point Likert scale:

Table 8.3 Risk perceptions

Variable	Variable description	Mean (SD)
DangerA	Dummy, 1=Subject perceives consumption of agricultural products from region A as being dangerous to her health	0.864 (0.345)
NotDangerB	Dummy, 1=Subject perceives consumption of agricultural products from region B not being dangerous to her health	0.830 (0.378)

I perceive consumption of agricultural products from region A as being dangerous for my health

I perceive consumption of agricultural products from region B not being dangerous for my health

Replies were then dummy-coded for the subsequent analysis with one corresponding to agreement (declaring higher than 5 in the original scale) and 0 otherwise. Table 8.3 shows the mean scores. Results suggest that 86 % of the respondents perceive significant risks to be entailed in consumption of agricultural products from area A that bears similar characteristics to the Asopos RB. This was true although significant effort had been placed to accurately present the scientific information available on the ecological status and pollution loads in the basin. In particular respondents were reminded the lack of epidemiological data that could establish causality between food consumption from the area and health symptoms. Accordingly, consumption of products from region B is considered safe from the overwhelming majority. To examine the degree to which risk perceptions can influence real behavior we introduce the risk variables in the regression analysis. Results can have useful implications for risk communication and can feedback in the developing awareness raising material.

4.2 Econometric Results

To account for the panel nature of our data, we estimated a random effects regression model. The bid function is specified as a function of the socioeconomic characteristics of the respondents, the initial monetary endowment, risk perceptions and potato consumption habits. We further control for the auction round to account for potential learning effects. The regression function is thus specified as:

$$
\begin{aligned}
&Bid_{it} = \\
&\left(\begin{aligned}
&b_0 + b_1 Charity_{it} + b_2 HealthRisk_{it} + b_3 TotFee_{it} + b_4 T_{2,it} + b_5 T_{3,it} + b_6 T_{4,it} + b_7 T_{5,it} \\
&+ b_8 Age_{it} + b_9 Gender_{it} + b_{10} Income_{2,it} + b_{11} Kids_{it} + b_{12} Educ_{2,it} + b_{13} DangerA_{it} \\
&+ b_{14} NotDangerB_{it} + b_{15} PotatoConsumption_{2,it} + b_{16} PotatoConsumption_{3,it} \\
&+ b_{17} PotatoConsumption_{4,it}
\end{aligned}\right) \\
&+ e_{it} + u_i \qquad (8.1)
\end{aligned}
$$

Table 8.4 Econometric results

Variable	Coef. (St Error)
Constant	1.769 (1.267)
Charity	–0.251 (0.155)
HealthRisk	–0.208 (0.148)
TotFee	–0.073 (0.050)
T_2	0.058* (0.033)
T_3	0.149*** (0.033)
T_4	0.189*** (0.033)
T_5	0.236*** (0.033)
Age	0.005 (0.008)
Gender	–0.094 (0.171)
$Income_2$	0.235 (0.146)
$Educ_2$	0.007 (0.155)
Kids	–0.068 (0.168)
DangerA	0.079 (0.236)
NotDangerB	0.436** (0.209)
$PotatoConsumption_2$	0.392 (0.275)
$PotatoConsumption_3$	–0.170 (0.241)
$PotatoConsumption_4$	–0.004 (0.251)
R-squared	0.278

* p<.1; ** p<.05; *** p<.01

Table 8.4 shows the coefficient estimates from the regression. Results show that consumers are indeed WTP a premium to exchange their agricultural endowment. Mean bid is 60 cents per kilo of potatoes. Bids are increasing across rounds. As expected subjects that perceived consumption of agricultural products from region B to pose no health risk, bid 0.43€ more to exchange their endowed products. Gender, income, households with underage kids and education do not seem to influence bidding behavior in a statistically significant way. Consumption habits of potatoes also do not have an effect on bidding behavior. Similarly, provision of extra information regarding the health risks of children was not statistically significant. Contrary to expectations, information effects are not evident in this experiment although commonly reported in the relevant literature. We contend that health risks for children are already considered when consumers state their value to hedge against health risks. This is an interesting avenue for future research as well.

5 Conclusions and Policy Recommendations

This chapter reports the results of an experimental auction designed to examine whether consumers are WTP a price premium to reduce the potential health risks associated with consumption of products that are cultivated in river basins that are not in a good ecological status. Health risks are a significant element of the environmental cost associated with water quality degradation and thus should be considered in the calculation of the total economic cost of water uses as the Directive requires.

To fulfill our research aims we endowed subjects with a kilo of potatoes cultivated in a river basin with characteristics similar to Asopos and gave them the opportunity to bid to exchange them with a kilo of similar variety potatoes from a river basin described as being in a good ecological status. Information on the advanced risks for children was also provided to half of our sample to assess the effect of information on valuation. Moreover, risks perceptions were also elicited at the end of the study and were accounted for in the analysis. Results of this study suggest that subjects are willing to pay more to upgrade from agricultural products cultivated in a polluted river basin products from areas in good ecological status. Mean upgrade bid is 60 cents per kilo of potatoes. This was so even though individuals were informed that epidemiological data that would allow assessment of the presence of risks for human health were not available.

This represents an external cost to the farmers in the area limiting their opportunity to market their products. It is indicative that although every caution was taken to accurately present the situation regarding the health implications of the consumption of agricultural products from region A, 86 % of the consumers perceived consumption of the products as being dangerous for their health. Economic instruments are thus in order, as the WFD requires, to compensate farmers for the income loss they bear if the analysis of water uses confirms that agriculture is not contributing to the ecological degradation. Polluters should then be identified and charged so that the total cost of water services, including external costs of ecosystem degradation, is recovered according to the polluter pays principle that the WFD introduces.

If on the other hand, epidemiological data fail to confirm causality between consumption of agricultural products from the Asopos RB and health risks, educational and information programs and campaigns should be developed so that consumers regain acceptance of the areas agricultural products. Past experience has shown that such campaigns are effective in diminishing consumers' resistance (Nayga 1996).

Appendix: Environmental Health Risk information

A.1 Environmental Profile of Region A

Region A is characterized by intensive industrial activity, with many of the industries not fulfilling the safety standards, and intensive agricultural activity. Underground water analysis has revealed the presence of heavy metals, such as chromium and nickel, which may have contaminated plants through irrigation. The severity of these substances for human health depends on the degree and the duration of the exposure. However, an epidemiological study assessing accurately the risks for human health from the consumption of agricultural products from region A, has not been performed yet. In addition, with respect to potatoes heavy metals tend to accumulate in the skin of potatoes and not in the interior that is commonly consumed.

A.2 Environmental Profile of Region B

Region B is classified as in good ecological status, according to the European Water Framework Directive. The good ecological status guarantees that pollution loads are minor such that there is no risk for human health and aquatic life. The agricultural sector follows good agricultural and environmental practices and there is no industrial activity in the area. Measurements in potatoes from the area revealed that the accumulation in heavy metals is far below the international safety levels.

A.3 Information on the Advanced Risks for Children

Children are extremely vulnerable to toxic substances in agricultural products. Children have a longer life span compared to adults and therefore are more likely to experience long-term effects from heavy metal contamination since exposure is cumulative.

References

Aldrich, L. (1994). Food safety policy: Balancing risk and costs. *Food Review, the magazine of food economics,* May–August 1994, pp. 9–13. U.S.D.A., Economic Research Service.

Commission of the European Communities (CEC). (2000). *Directive of the European Parliament and of the Council establishing a framework for Community action in the field of water policy.* 1997/0067 (COD), C5-0347/00.

Costa-front, G., Gil, G. M., & Traill, W. B. (2008). Consumer acceptance, valuation of and attitudes towards genetically modified food: Review and implications for food policy. *Food Policy, 33*, 99–111.

Florax, R. J., Travisi, C., & Nijkamp, P. (2005). A meta-analysis of the willingness to pay for reductions in pesticide risk exposure. *European Review of Agricultural Economics, 32*, 441–467.

Frewer, L. J., Howard, C., Hedderley, D., & Shepherd, R. (1998). Methodological approaches to assessing risk perceptions associated with food related hazards. *Risk Analysis, 18*, 95–102.

Lusk, J. L., & Shogren, J. F. (2007). *Experimental auctions: Methods and applications in economic and marketing research.* Cambridge, UK: Cambridge University Press.

Lusk, J. L., Jamal, M., Kurlander, L., Roucan, M., & Taulman, L. (2005). A meta-analysis of genetically modified food valuation studies. *Journal of Agricultural and Resource Economics, 30*, 28–44.

Morkbak, M. R., Christensen, T., & Gyrd-Hansen, D. (2008). Valuation of food safety in meat – A review of stated preference studies. *Food Economics - Acta Agriculturae Scandinavica, Section C, 5*, 63–74.

Nayga, R. M., Jr. (1996). Determinants of consumers' use of nutritional information on food packages. *Journal of Agricultural and Applied Economics, 28*(2), 303–312.

Rozan, A., Stenger, A., & Willinger, M. (2004). Willingness-to-pay for food safety: an experimental investigation of quality certification on bidding behaviour. *European Review of Agricultural Economics, 31*, 409–425.

Slovic, P. (1987). Perception of risk. *Science, 236*(4799), 280–285.

Stenger, A. (2000). Experimental valuation of food safety, application to sewage sludge. *Food Policy, 25*, 211–218.

Chapter 9
An Economically Efficient, Environmentally Sustainable and Socially Equitable Decision Support System for Asopos River Basin: A Manual of Measures

C. Doulgeris, S. Katsavouni, and D. Papadimos

Abstract This Chapter focuses on the development of a Decision Support System (DSS) for Asopos River Basin (RB) in order to achieve the holistic management of water resources and their ecosystems in the catchment area. As a result, the final output of this application is to identify measures that will contribute to the restoration of the ecosystem in Asopos River. The chapter at a first stage introduces the steps involved in the development of the DSS. Then the deliverables of this process are reported. Important components of the DSS include the gathering of available studies in the area, creation of a database of meteorological and hydrological parameters in the area, estimation in space and time of the parameters of the hydrological table and identification of pressures related to water resources in the area. Further components include the valuation of functions in the catchment, building of different scenarios conditional on changes on precipitation, temperature and water use, estimation in space and time of changes on the parameters of the hydrological table and pressures under the different scenarios. Finally, the DSS suggests measures for the economically efficient, environmentally sustainable and socially equal management of the basin. The considered measures involve among others a number of economic instruments such as standards and quotas, water abstraction taxes, pollution taxes, subsidies, tradable permits, voluntary agreements and liability legislation. It is expected that these instruments can provide the appropriate incentives for efficient water resources management.

Keywords Decision Support System • Basin management • Pressures on Aquatic ecosystems • Hydrological simulations • Ecosystem restoration

---

C. Doulgeris (✉) • S. Katsavouni • D. Papadimos
The Goulandris Natural History Museum,
Greek Biotope/Wetland Centre (EKBY), Thessaloniki, Greece
e-mail: haris@ekby.gr; sotiria@ekby.gr; papadimos@ekby.gr

P. Koundouri and N.A. Papandreou (eds.), *Water Resources Management Sustaining Socio-Economic Welfare*, Global Issues in Water Policy 7,
DOI 10.1007/978-94-007-7636-4_9,

1 Introduction

The Asopos River, which drains an area of 719 km^2, rises on Mount Lefktra, on the northern flank of the Kithairon range, collects water from tributaries in the southern part of the Theban lowlands, and continues eastward to flow into the South Gulf of Euboea near Oropos. This terrain has a particularly regular relief, with the highest ground in the southern part of the river basin; the average elevation is 357 m, and the highest point 1305 m. Land uses in the region include arable farming, especially in the upper and middle course of the river, forest and scrubland in the southern part of the basin, and a fair amount of industry.

Intense human activity, both agricultural and industrial, has developed in the river valley. Over its entire length the Asopos runs through farmed land, while the Avlona – Inofita – Inoi – Schimatari zone is heavily industrialized. As a result, the Asopos frequently receives wastes and effluents of varied composition and origin, particularly in its middle course (Giannoulopoulos 2008; Bakos 2009), the principle pollutants being hexavalent chromium, nitrates, divalent lead and chlorine ions, which are present in concentrations in excess of the parametric values established in Community law on the quality of water for human consumption.

The main water systems in the Asopos RB are the channel and mouth of Asopos River and the Oropos lagoon, which is the sole remaining coastal wetland in the area. The wetlands complex of the mouth of the Asopos and the Oropos lagoon comprises characteristic types of coastal wetlands, creating a variety of habitats and a striking physiognomic mosaic. The change in water depth inwards from the shore resulting from the local microtopography and periodic seasonal flooding during the year creates a variety of water depths favoring the appearance of many types of vegetation and species of flora and fauna.

The diversity of wildlife occurring in the region is reflected in the 120 species of birds that have been recorded in the study area (Tzali et al. 2009, Dimaki 2011 In Katsavouni et al. 2012), more than in any other wetland in Attica, including raptors, wading birds, overwintering waterfowl, and shorebirds that are found year round but in greater numbers in the winter and during the migrating season. There are also gulls, most notably the Mediterranean gull (*Larus melanocephalus*), which is in the Red Data Book of threatened species, and the little tern (*Sterna albifrons*), which nests in the area, and several species of perching birds. Forty-nine of the avifauna species found here are listed in Annex I to the Birds Directive. With regard to other fauna species, amphibians and reptiles observed in the region include the green toad (*Pseudepidalea viridis*), the Balkan water frog (*Pelophylax kurtmuelleri*), the margined tortoise (*Testudo marginata*), four lizard species [i.e. the European copper skink (*Ablepharus kitaibelii*), the ocellated skink (*Chalcides ocellatus*), the Turkish gecko (*Hemidactylus turcicus*) and the Balkan green lizard (*Lacerta trilineata*)] and the Montpellier snake (*Malpolon monspessulanus*). Certain species of mammals have also been observed in the region, such as the black rat (*Rattus rattus*) and five species of bats [i.e. the Nathusius' Pipistrelle (*Pipistrellus nathusii*), the Kuhl's Pipistrelle (*Pipistrellus kuhlii*), the Savi's Pipistrelle (*Hypsugo savii*), the Geoffroy's

Bat (*Myotis emarginatus*), and the Brandt's Bat (*Myotis brandtii)*]. The habitats for most species of reptiles and mammals are severely degraded.

2 Surface and Groundwater

The hydrographic network of the Asopos is not particularly well developed, since the extensive areas of highly karstified limestone allow the water in the torrents to percolate into the groundwater system, resulting in an uneven ramification on the two sides of the channel. Formerly, even the Asopos itself held water only for a very short period, despite its relatively large catchment area, because of the rapid percolation of surface water into the aquifer. Today, because of the effluents entering the river, parts of the Asopos have water even in the summer months.

The streams in the river basin do not have a permanent flow, and only in certain areas with impermeable formations (schists, clay deposits) are there small torrents that retain a flow of water for a certain period of time: examples include the Lantikos and the Gouras, which traverse the neopaleozoic schists north of Platy Vouno. The Liveas, northwest of Malakasa, whose course runs through the quaternary clays that drain the region, has a seasonal flow. The longest streams in the north part of the river basin are the Sklirorrema, the Potisiona and the Vathi, which drain most of its north side. The streams in the south side of the basin, the Xerias, Bresiko, Kalamata, Lykorrema and Aghios Dimitrios, drain the karst basins on the north face of Mount Pastra and the eastern slopes of the Kithairon (Bakos 2009; Tsarabaris 2010).

No systematic measurements of the discharge of the watercourses in the river basin have been made. A study carried out by Frangopoulos et al. (1992) for the Ministry of Agriculture calculates the discharge volume in the Asopos basin at 26×10^6 m^3/year; this is based on limited-time observations and estimates drawn from bibliographical data, and cannot be considered representative (Bakos 2009). The mean annual discharge rate at Rapentoza over the period 1947–1956 was 0.73 m^3/s (Economou et al. 2001), while for the whole of the Asopos basin the total annual discharge at the river mouth is estimated at 70.2×10^6 m^3.

The bedrock in the area is represented by thick formations of Mesozoic limestones, tertiary and quaternary formations and deposits, and limited areas of Paleozoic schists. The tertiary formations comprise limestones, marls, sandstones and conglomerates, and the quaternary of diluvial formations, debris and alluvial deposits (Tsarabaris 2010). In the western part of the Asopos basin the tertiary and quaternary formations and deposits are very deep, with a thickness ranging from 150 to 300 m (Frangopoulos et al. 1992). These formations overlie a karstic aquifer, which is fed from the north slopes of Mount Parnes and from surface limestone outcrops within the Asopos basin.

In the vicinity of Aghios Thomas, the groundwater level dropped by an average of 22.5 m between 1951 and 1976 (Bakos 2009), while the pumping discharge from the boreholes was reduced by 60 %. Currently there are a large number of boreholes in the river basin, with an average discharge of 30–50 m^3/h, which draw water from

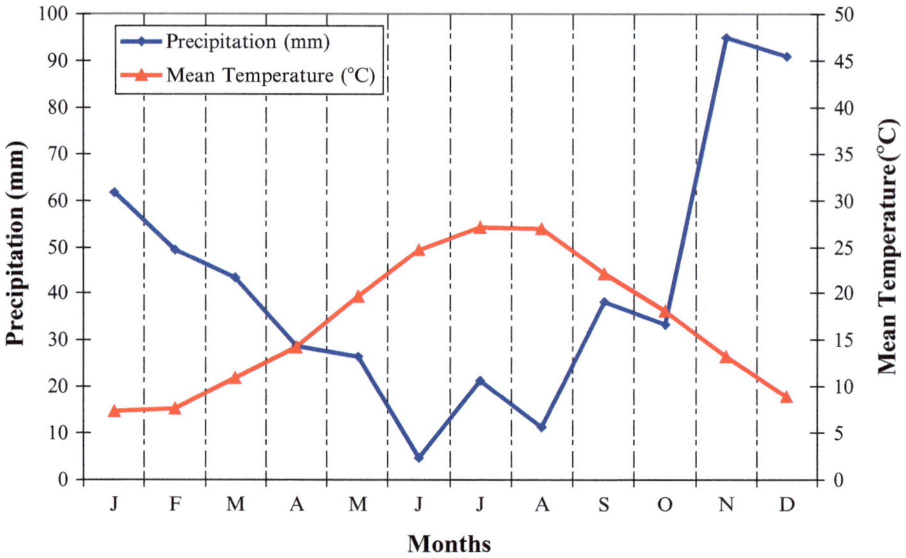

Fig. 9.1 Ombrothermic diagram in the Asopos RB

a depth of 150–160 m below ground level. The average distance between them is 150–200 m, while each borehole serves an area of approximately 12 ha.

3 Meteorological Data

Data from the meteorological stations at Kallithea, Tanagra and Marathon for the period October 1999 to September 2010 show annual average precipitation levels of 534.5 mm, 502.9 mm and 625 mm respectively. The mean annual air temperature at the Tanagra and Marathon stations is, respectively, 16.7 °C and 17.5 °C.

Based on Lang's rainfall index (Trewartha and Horn 1980), which expresses the ratio of average monthly precipitation to average monthly temperature, the climate is hyper-arid from May through September and varies from arid to wet over the rest of the year. Figure 9.1 gives the ombrothermic diagram at the Tanagra meteorological station, based on average monthly precipitation (mm) compared to the corresponding average monthly temperature (°C).

4 Evaluation of Wetland Functions and Values

Wetlands are considered invaluable natural, economic and social assets and efforts are made both to protect them and to restore their functions and values. As it relates to the functions, it is widely recognized that they can perform various, such as: (a) to store water, (b) to enrich the groundwater aquifers, (c) to change the flood peaks, (d) to transform and remove nutrients and (e) to support the food webs. Wetland functions are physical, chemical and biological processes that they perform. Values for humans result from the structural features and functions of wetlands. These are the goods and services provided or potentially provided to humans as a result of functions that take place there.

The identification and assessment of degraded or under-degradation functions and values, is the first step in defining management objectives and the subsequent adoption of appropriate management measures to achieve them. Through appropriate interventions in wetland functions, degraded values are restored or deterioration is prevented of those values that are in danger directly or indirectly because of the declining trend in their respective functions.

A full assessment of functions and values has been undertaken on the basis of existing methods, such as Wetland Evaluation Technique-WET (Adamus et al. 1987) and EVALUWET (Maltby 2009). According to the methods, the wetland is separated into discrete hydrogeomorphological units based on topographic criteria, hydrology, soil, etc. and the functions in each of these units are evaluated (Brinson 1993a, b; Maltby 2009). In the context of the current work, we attempted identification and assessment of the functions and values of the wetlands in Asopos basin. These wetlands are: the Asopos riverbed, including its estuary, and the lagoon of Oropos. Within this approach, we assessed the ability of wetlands to perform certain functions, according to their structure and special features and those of the river basin (Marble 1992). The assessment area was divided into hydrogeomorphological units according to the method EVALUWET and for assessing functions and values we applied the method of WET. According to the method EVALUWET as *hydrogeomorphological unit* is defined part of the landscape with uniform morphology and consistent hydrological regime.

Identified and delineated in the study area were the following 10 hydrogeomorphological units (HGMUs): A. Western part of Asopos riverbed. B. Central part of Asopos riverbed. C. Eastern part of Asopos riverbed D. Estuary of Asopos. E_{1-4}. Four regions located west and east of the estuary of Asopos and in the vicinity of Oropos lagoon. Z. Oropos lagoon. H. South and east of Oropos lagoon.

The functions that were considered necessary to evaluate in the area were: (a) water storage, (b) food web support: aquatic life and avifauna, (c) nutrients removal and transformation, (d) sediment and toxic trapping, (e) floodwater attenuation, (f) groundwater recharge, (g) shoreline stabilization. The degree of performance of each function was qualitatively assessed and given the designation “high”, “moderate”, “low” or “none”. In summary, the degree of performance of wetland functions in each unit is given in Table 9.1.

In the case of study area, the following wetland values were evaluated: (a) biological, (b) scientific, (c) education (d) recreational, (e) hunting, (f) improving of water quality, (g) protection against floods, (h) protection against erosion. The evaluation was done with consideration of the degree of performance of wetland functions and the status quo, with respect to services and goods that can accrue to the man from the wetland. The evaluation leads to characterize the degree of expression of each value as a “high”, “moderate”, “low” or “none”. Below is the assessment of values, which occurred in regions: (a) Asopos riverbed (HGM includes units A, B and C in their entirety), (b) estuaries (including the HGM units D and E_{1-4} entirely), and (c) lagoon of Oropos (HGM includes units Z and H in their entirety) (Table 9.2).

Table 9.1 Performance degree of the wetland functions in the study area

	Wetland functions							
	Water storage	Food web support: aquatic life	Food web support: avifauna	Nutrient removal and transformation	Sediment and toxic trapping	Floodwater attenuation	Groundwater recharge	Shoreline stabilization
HGM unit	Performance degree							
A	Low	Low	Low	Low	High	Low	High	None
B	Moderate	Low	Low	Low	High	Low	Moderate	None
C	Low	None	Low	Low	High	Low	High	None
D	Low	Low	High	Low	High	Low	Moderate	High
E_{1-4}	Low	Low	Moderate	Moderate	Moderate	Moderate	Low	High
Z	Low	Low	High	Low	Moderate	Moderate	Low	High
H	None	Low	Moderate	Low	Moderate	Moderate	Low	High

Table 9.2 Performance degree of the wetland values in the study area

	Wetland values							
	Biological	Scientific	Educational	Recreational	Hunting	Improving of water quality	Protection against floods	Protection against erosion
HGM unit	Performance degree							
Asopos riverbed (HGMUs A, B, C)	Low	Low	Low	None	Moderate	Low	Low	None
Asopos estuary (HGMUs D, E_{1-4})	Moderate	Moderate	Moderate	Low	Moderate	Moderate	Moderate	Moderate
Oropos Lagoon (HGMUs Z, H)	Moderate	Moderate	Moderate	Moderate	Moderate	Moderate	Moderate	Moderate

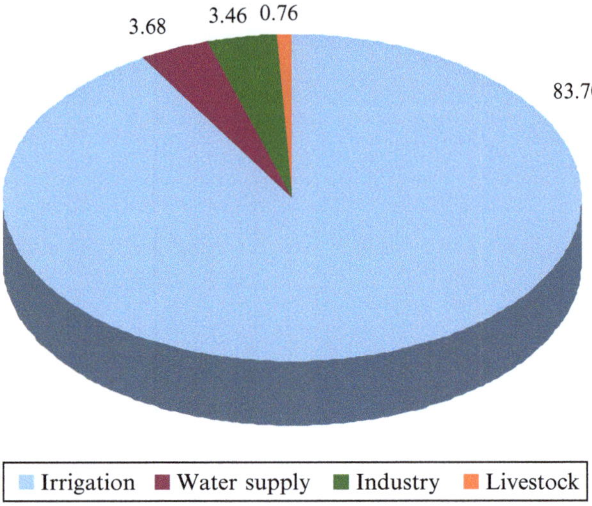

Fig. 9.2 Annual water demand by users in the Asopos RB (in million m^3)

5 Determining Pressures on the Aquatic Ecosystems

Industrial activity in the Asopos RB severely degrades the quality of the river's water, while the human impact on the coastal wetlands is also very strong, chiefly from residential pressures, vehicle traffic in the summer months as bathers use the beaches, and the discarding and accumulation of rubbish. Road construction and the expansion of residential development towards the wetlands exert pressure on the Oropos lagoon. A multitude of roads form an urban grid and the pressure of residential development extends right to the shoreline. Taking all this into account, it is obvious that the wetlands complex is subject to continuous degradation, due mainly to the expansion of residential construction at the expense of the wetlands region and the pollution of the Asopos River. Moreover, evaluation of wetlands functions and values has led to recognition of the degradation of the wetlands ecosystems and the need to rehabilitate them.

Meeting the water needs generated by urban water supply, irrigation, livestock and industry exerts further pressure on the aquatic ecosystems. It is clear from the graph in Fig. 9.2, which charts water consumption for the above uses, that irrigation is by far the heaviest user (91 %).

6 Hydrological Simulation of the Asopos River Basin

6.1 Setting up an Integrated Hydrological Model

An integrated hydrological model of groundwater and surface water in the Asopos RB was developed by setting up the MIKE SHE and MIKE 11 modeling systems.

The hydrological model of the Asopos RB is part of the decision support system, and its object is to simulate the availability of water resources in order to assess the impact of pressures exerted on them by water users.

In MIKE SHE/MIKE 11 the processes of the hydrological cycle in the Asopos RB are simulated using the modules outlined in Table 9.3. The region for which the simulation system was developed is the lowlands part of the basin (Fig. 9.3), where the human-induced pressures on the aquatic ecosystems were identified.

The simulation used – *inter alia* – meteorological, morphological, pedological, geological and land-use data. Rainfall amounts and intensities are the driving force behind almost all the processes of the hydrological cycle, and the Asopos RB receives most of its rainfall (74.1 %) in the 6 months from October to March, and just 25.9 % between April and September.

The relief of the ground determines both the drainage areas and those of surface runoff, and also shapes the natural upper limit of the unsaturated and – albeit in certain conditions – the saturated zone. As regards the spatial distribution of the various land uses, these are classified as agricultural, forest or urban before being entered into the model (Fig. 9.4). The quantity of water needed for irrigation puts heavy pressure on the groundwater and is spatially distributed over the lowlands area of the river basin before being included in the model.

Table 9.3 Processes of the hydrological cycle simulated in the Asopos RB

Process of hydrological cycle	Module	Combination of modules and systems	Dimensionality of the simulation	Equations describing the process
Overland flow and storage	MIKE SHE OL	MIKE SHE SZ, UZ AND MIKE 11	1 – D	Routing overland flow based on the Manning equation and empirical formulas
Movement of water in open channels (discharge and flow depth)	MIKE 11 HD	MIKE SHE SZ, OL	1 – D	Saint Venant equation
Infiltration and storage of water in the unsaturated zone, saturated zone recharge	MIKE SHE UZ	MIKE SHE SZ, OL	1 – D	Two-layer water balance method
Evapotranspiration	MIKE SHE ET	MIKE SHE UZ, OL	–	Kristensen & Jensen
Movement and storage of water in aquifers	MIKE SHE SZ	MIKE SHE UZ, OL AND MIKE 11	2 – D	Groundwater flow based on the Darcy equation

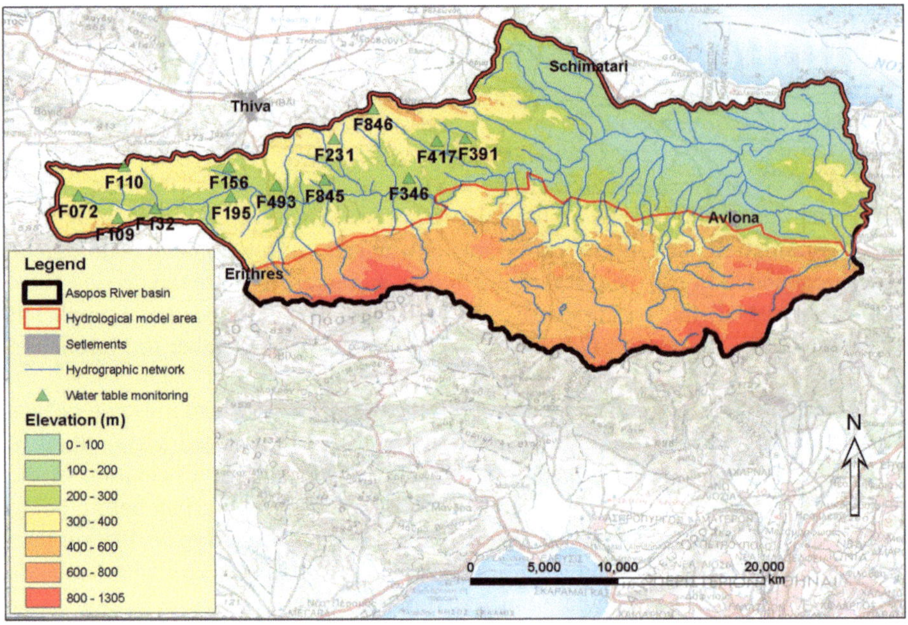

Fig. 9.3 Hydrological model area in the lowlands of Asopos RB

6.1.1 Hydrological Processes in the MIKE SHE

The hydrological processes covered by the Asopos RB model include overland flow, evapotranspiration, infiltration into the unsaturated zone and flow in the aquifers.

Overland flow occurs when there is heavy rainfall and the rate of precipitation exceeds the rate of absorption into the soil. The course and the final quantity of water in surface streams are determined by the relief and the roughness of the ground and by losses due to evapotranspiration and percolation. This process is simulated by means of the numerical solution of the continuity equation, using as auxiliary equations an empirical relationship between flow depth and ground gradient coupled with Manning's equation (Crawford and Linsley 1966).

Water flow in the unsaturated zone is simulated using a simple two-layer water balance approach, where the first layer extends from the surface of the ground to the root zone and the second from the bottom of the root zone to the groundwater table (or the ceiling of the saturated zone). In each layer the humidity conditions are considered as uniform in terms of depth. The main object of this approach is not a detailed simulation of the movement of the water in the unsaturated zone, but an estimate of real evapotranspiration, the quantity of water that is held in the unsaturated zone, and the quantity of water that percolates into the saturated zone (Yan and Smith 1994).

Evapotranspiration is calculated taking account of the following processes:

1. Part of the precipitation is intercepted by the leaf canopy, and evaporates.
2. The rest reaches the surface of the ground, where it either runs off or penetrates into the unsaturated zone.

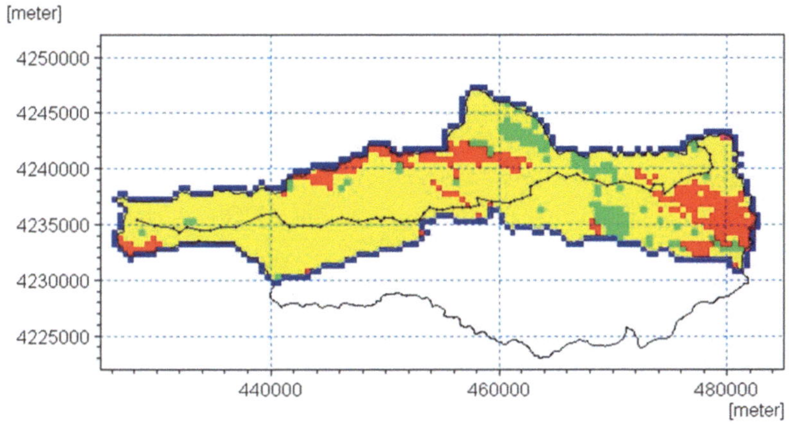

Fig. 9.4 Land use in the model area of Asopos RB (*yellow*: agricultural land, *red*: forest, *green*: urban and industrial areas)

3. The quantity of water that is stored in the unsaturated zone (the root zone) either evaporates or is transpired by the vegetation.
4. The remaining quantity of water that infiltrates into the soil is considered as recharging the aquifers.

The soil infiltration rate determines the ratio of surface runoff to the water percolating into the unsaturated zone; the lower the infiltration rate, the greater the quantity of water that is available for surface runoff, with a corresponding reduction in the quantity of water available to percolate into the unsaturated zone.

Water flow in the saturated zone is one of the hydrological processes of the integrated model of surface and groundwater that interacts with all the others – overland flow, flow in the unsaturated zone, evapotranspiration, stream flow – and uses them in boundary conditions.

Groundwater movement is simulated by the numerical solution of the three-dimensional differential equation:

$$\frac{\partial}{\partial x}\left(K_{xx}\frac{\partial h}{\partial x}\right)+\frac{\partial}{\partial y}\left(K_{xx}\frac{\partial h}{\partial y}\right)+\frac{\partial}{\partial z}\left(K_{xx}\frac{\partial h}{\partial z}\right)-Q=S\frac{\partial h}{\partial t} \tag{9.1}$$

where h is the piezometric head on the aquifer (m), K_{xx}, K_{yy}, K_{zz} the hydraulic conductivity along the x, y, z axes (m/s), Q the volume of water that is added or abstracted per unit volume of the aquifer (s^{-1}), S the specific storage coefficient, x, y, z the axes of the Cartesian coordinate system, and t is time.

6.1.2 Interaction Between Surface- and Groundwater

The model takes account of the interaction between the groundwater and the Asopos River by coupling MIKE SHE with MIKE 11. Water movement in the Asopos is simulated in the MIKE 11 environment by the solution of the Saint Venant equations:

$$\frac{\partial Q}{\partial x} + \frac{\partial A}{\partial t} = q \tag{9.2}$$

$$\frac{\partial Q}{\partial t} + \frac{\partial}{\partial x}\left(\alpha \frac{Q^2}{A}\right) + gA\frac{\partial h}{\partial x} + \frac{n^2 gQ|Q|}{AR^{4/3}} = 0 \tag{9.3}$$

where Q is the discharge of the river (m^3/s), A the cross-section (flow) area (m^2), q the lateral inflow into the river (m^2/s), h the water level above the reference datum (m), x the longitudinal direction of flow (m), t time (s), n Manning's coefficient of friction ($s/m^{1/3}$), R the hydraulic radius (m), g the coefficient of gravitational acceleration (m^2/s) and α the coefficient of velocity distribution (–). Depth of flow and discharge along the course of the Asopos are calculated by means of the solution of Eqs. 9.2 and 9.3 in a one-dimensional computational network of an implicit finite difference scheme.

The Asopos River was input to MIKE 11 taking account of the digital model of the relief of the basin used by MIKE SHE, permitting accurate hydraulic communication between MIKE SHE and MIKE 11. As the river flows through its drainage basin, it interacts with the groundwater. This interaction is bi-directional (Fig. 9.5), and groundwater recharge from the river (Q_2) and drainage from the groundwater into the river (Q_1) are calculated by means of a formula based on Darcy's law.

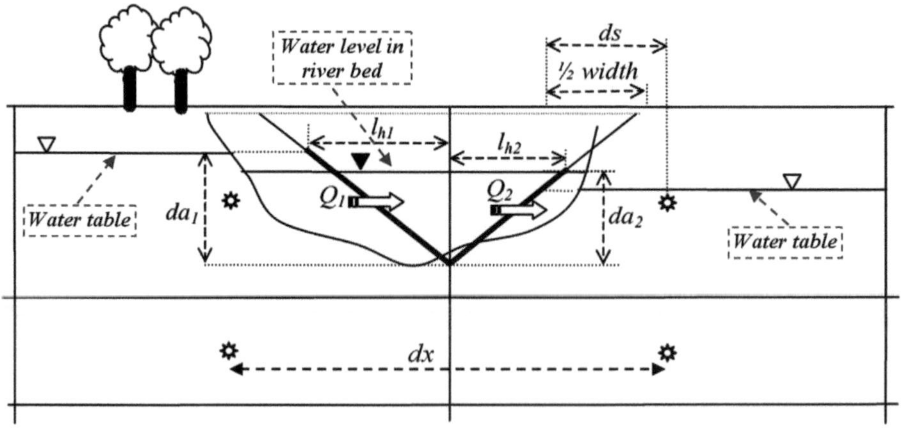

Fig. 9.5 Interaction of surface- and groundwater in a hypothetical river cross section (Source: DHI (2009))

6.2 Calibration of the Hydrological Model

The structure of the distributed hydrological simulation systems permits spatial variation of the characteristics of the basin in a network of points in the form of a rectangular grid. Frequently, the application of models to a river basin requires several thousand points, each of which is characterized by different parameters and variables. A distributed hydrological simulation system like MIKE SHE has the potential to deliver a large number of parameters, the values of which must be determined when calibrating the model. Some of these parameters can be estimated from field data, e.g. vegetation distribution charts, geological and soil sections, etc.

In calibrating the model of a river basin, the available field data should be used to determine spatial units, in each of which there is little fluctuation in the value of a specific model parameter or group of parameters. In the Asopos RB, the parameterization of the model took into account the following spatial units:

- Categories of land use, for which the leaf area index and depth of root zone are determined.
- Categories of soil type, for which the humidity at field capacity, the humidity at permanent wilting point and the hydraulic conductivity at saturation are determined.
- Areas where the hydraulic parameters determined are those relating to overland flow and flow in the saturated zone.

On the basis of the results of the model for flow in the saturated zone, the parameters of hydraulic conductivity K and specific yield S were calibrated using the measured values of the groundwater table at the monitoring points. During the calibration three sets of values were tested for hydraulic conductivity (K1, K2, K3) and three for specific yield (S1, S2, S3). From the adjustment of the model it was determined that hydraulic conductivity ranges from 10^{-3} m/s to 10^{-9} m/s, and specific water yield from 0.13 to 0.17.

The mean absolute error (MAE), the root mean squared error (RMSE) and the correlation coefficient (r) are the statistical criteria that were used in checking the results of the model and comparing them to the measured values of the groundwater table (Table 9.4).

Figures 9.6, 9.7, and 9.8 show the variation in the calculated model values in relation to the measured values of the groundwater table at selected monitoring points (Fig. 9.3).

6.3 Analysis of the Water Balance Parameters

The surface runoff from the river basin is collected along the length of the river's course; Fig. 9.9 gives the calculated annual surface runoff and the surface runoff coefficient (the ratio of surface runoff to amount of precipitation) at the mouth of the

Table 9.4 Mean Absolute Error (MAE), Root Mean Squared Error (RMSE) and Correlation Coefficient (r) of the observed and calculated values of groundwater table in the Asopos RB

	Statistical criteria		
Monitoring stations of water table	MAE (m)	RMSE (m)	R
F072	6.74	7.79	0.58
F109	2.0	3.13	0.81
F132	2.18	3.01	0.91
F110	10.88	12.24	0.70
F156	1.65	1.83	0.92
F195	3.79	5.67	0.48
F493	1.91	2.07	0.82
F845	1.67	2.12	0.93
F231	2.29	2.77	–0.04
F846	3.62	4.67	0.85
F346	1.68	1.87	0.89
F417	1.84	2.22	0.47
F391	1.07	1.33	0.85

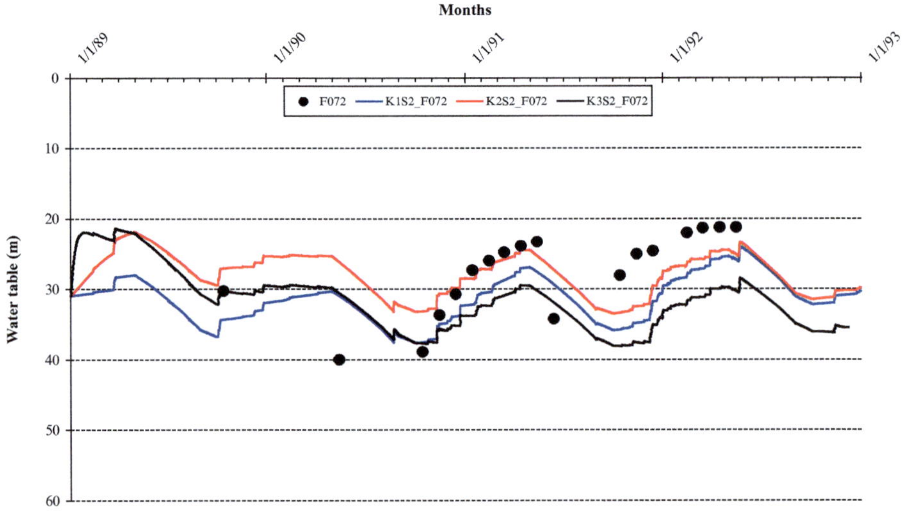

Fig. 9.6 Variation of observed and calculated (K1S2, K2S2, K3S2) values of groundwater table in the station F072

Asopos. The average value for the surface runoff coefficient is 0.2. Annual surface runoff A (mm) can be estimated from annual precipitation P (mm) based on the relationship A= 0.2665×P – 31.444 (correlation coefficient, R^2=0.93).

The water balance of surface water in the lowland part of the Asopos RB is given by the equation:

$$P + Irr = \mathrm{ET} + RF\_1 + \delta \tag{9.4}$$

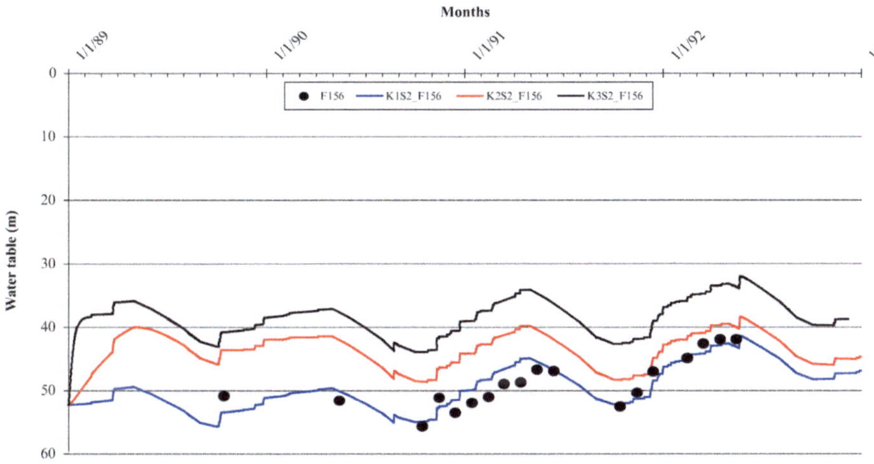

Fig. 9.7 Variation of observed and calculated (K1S2, K2S2, K3S2) values of groundwater table in the station F156

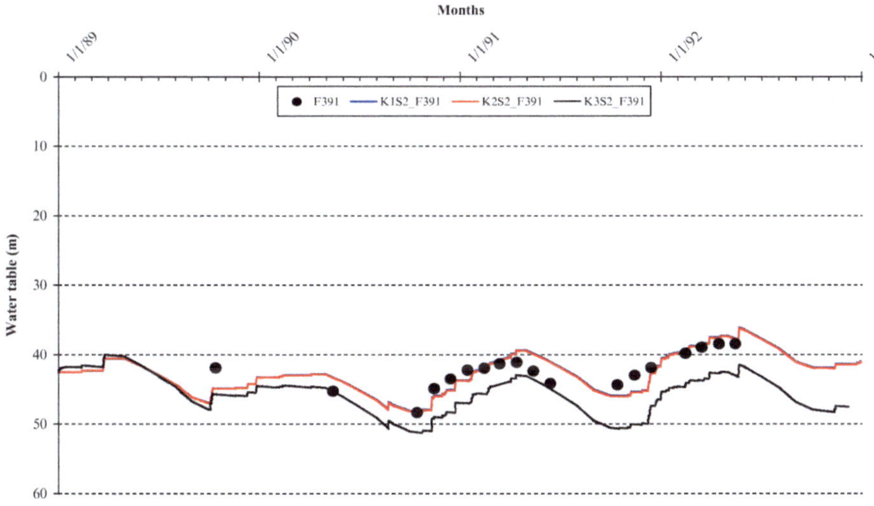

Fig. 9.8 Variation of observed and calculated (K1S2, K2S2, K3S2) values of groundwater table in the station F391

and the water balance for groundwater by the equation:

$$\delta + GW\_\mathrm{I} = Irr + RF\_2 + GW\_\mathrm{O} + GW\_\mathrm{S} \tag{9.5}$$

Combining Eqs. 9.4 and 9.5 gives the balance of surface and groundwater in the lowland part of the Asopos RB, which is expressed by the equation:

$$P + GW\_I = \mathrm{ET} + RF + GW\_O + GW\_S \tag{9.6}$$

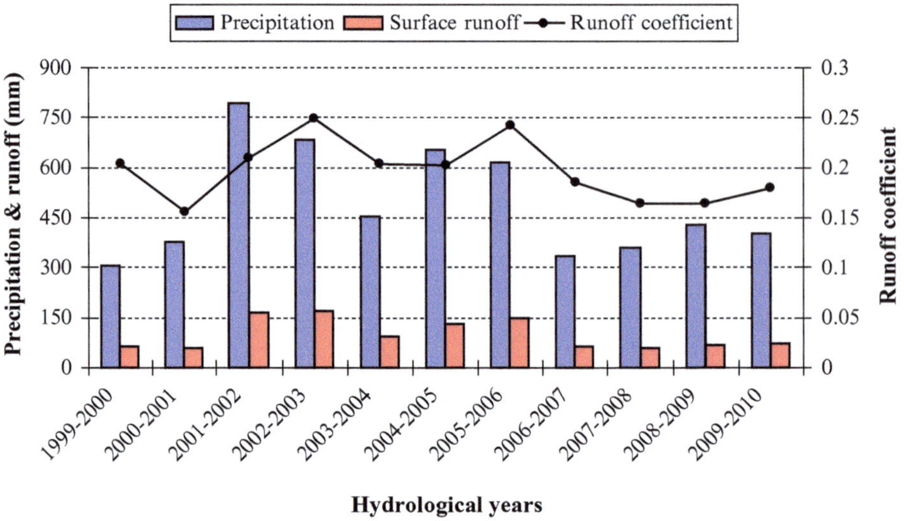

Fig. 9.9 Precipitation, surface runoff and runoff coefficient in the Asopos RB

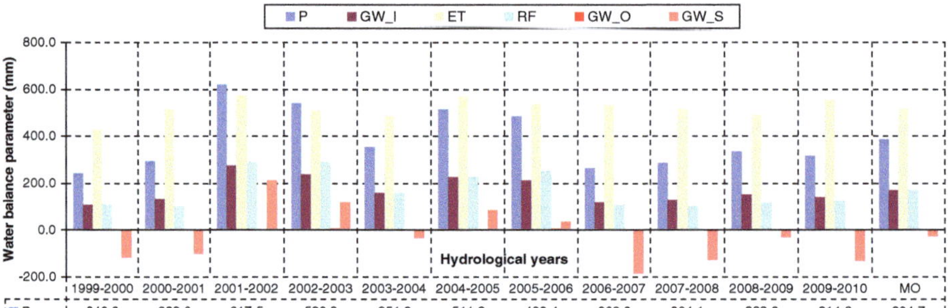

	1999-2000	2000-2001	2001-2002	2002-2003	2003-2004	2004-2005	2005-2006	2006-2007	2007-2008	2008-2009	2009-2010	MO
P	240.6	293.6	617.5	538.2	351.9	511.3	482.4	263.9	284.1	333.9	314.2	384.7
GW_I	107.0	131.5	274.7	236.1	156.9	226.4	212.3	116.6	126.7	149.1	141.0	170.7
ET	427.0	511.1	570.8	510.2	486.0	569.9	535.2	530.2	516.6	489.0	553.8	518.2
RF	105.9	100.1	287.1	288.3	158.1	226.7	252.4	104.9	101.7	118.8	123.8	169.8
GW_O	0.9	0.7	2.7	4.0	1.8	2.2	3.6	0.9	0.9	0.6	0.6	1.7
GW_S	-119.5	-103.7	208.9	116.8	-36.0	82.9	35.4	-184.4	-129.8	-32.5	-133.8	-26.9

Fig. 9.10 Hydrological balance in the lowlands of Asopos RB

where P is the amount of precipitation (mm), Irr the irrigation water requirements (mm), ET evapotranspiration (mm), δ the interaction between surface and groundwater (mm), RF_1 the overland flow from the basin to the river (mm), RF_2 the subsurface flow from the basin to the river (mm), GW_I the subsurface inflow from the mountainous part of the basin (mm), GW_O the subsurface outflow to the sea (mm), GW_S the change in the volume of groundwater (mm) and (RF = RF_1 + RF_2) the flow from the basin to the river. The quantity of water lost into the atmosphere from evapotranspiration includes the water used for irrigation.

The inflow of water into the basin is the sum of the amount of precipitation (P) and subsurface inflow (GW_I); it averages 662.8 mm/year. With a total outflow from the basin (ET + RF + GW_O) of 689.7 mm/year, the result is a net loss of groundwater amounting to 26.9 mm/year (Fig. 9.10).

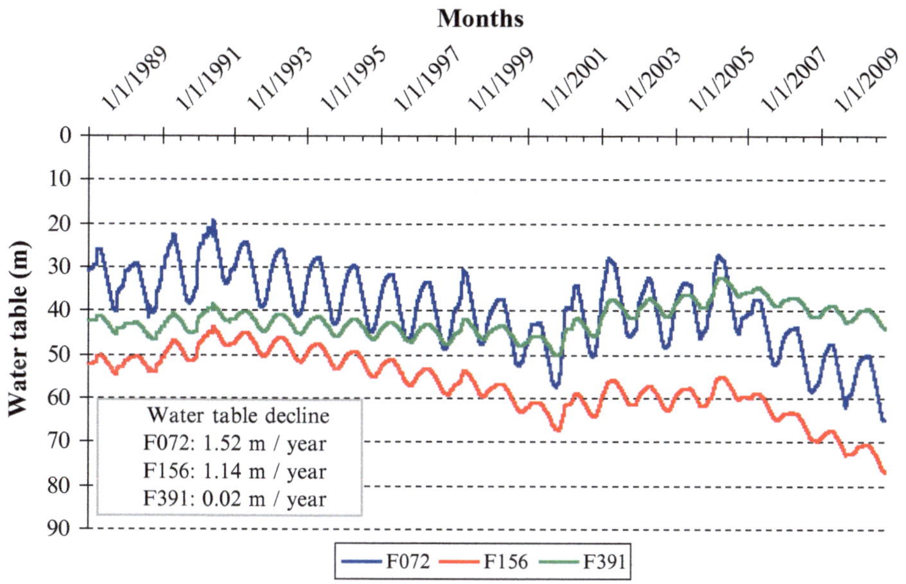

Fig. 9.11 Water table variation in the Asopos RB

The pressure on the Asopos RB caused by irrigation has led to a reduction in the volume of groundwater and hence a lowering of the water table. The magnitude of this pressure varies, with the heaviest pressures occurring in the western part of the basin, where larger quantities of water are required for irrigation. The fluctuation in groundwater levels has resulted in a continuous year-on-year decline in the water table (Fig. 9.11).

7 Proposals for Climate Change Adaptation Measures

7.1 Climate Change Scenarios and Projected Changes in Irrigation Water Requirements

The climate changes that have been observed and their effects on the human population and the ecosystems have created the need for a study of climate characteristics. Climate models are an attempt to reproduce the natural climate, and possible climate changes are studied on the basis of projected greenhouse gas emissions. The creation of these scenarios is based on the documented scientific opinion of the United Nations' Intergovernmental Panel on Climate Change (IPCC) that the climate changes that have been observed, such as rising temperatures, are due to increased emissions of greenhouse gases, and particularly carbon dioxide.

Table 9.5 Rainfall and temperature variation in the Asopos RB for the climate change scenario A1B

	Rainfall		Temperature	
Scenario/simulation period	mm	%	°C	%
1999–2010	502.9		16.7	
A1B/2020–2031	469.7	−6.6	18.4	10.1
A1B/2070–2081	398.3	−20.8	20.4	22.1

The IPCC's Third Report presents some 40 greenhouse emissions scenarios (Nakićenović et al. 2000), based on projected developments in population size, energy policies, economic growth rates and future advances in technology (Kapsomenakis et al. 2011).

The Academy of Athens' Research Centre for Atmospheric Physics and Climatology has developed databases and simulated models based on scenarios A2, A1B, B1 and B2. In the Asopos RB, the emissions scenario used for rainfall and temperature was A1B, which has the following characteristics (EMEKA 2011):

Scenario A1B: Rapid economic growth. Particularly intense energy use, but with a parallel spread of new and efficient technologies; use of both fossil fuels and alternative sources of energy; minor changes in land use. Rapid increase in the global population up to 2050 and gradual decrease thereafter. Marked increase in the concentration of CO_2 in the atmosphere, reaching 720 ppm by 2100.

The change in the amount of rainfall in the Asopos basin associated with greenhouse gas emissions scenario A1B was plotted for the periods 2020–2031 and 2070–2081, using 1999–2010 as the reference period. The reduction in annual rainfall for the decade 2070–2081 will be greater (−20.8%) than for the decade 2020–2031 (−6.6%) compared to the reference period 1999–2010. As for temperature change, the model predicts a rise of 10.1 % for the period 2020–2031 and of 22.1 % for the period 2070–2081 (Table 9.5).

Irrigation consist a significant pressure in Asopos RB and regards almost 91 % of total water requirements. Irrigation Water Requirements (IWR) were estimated based on climate data and applying the Blaney-Criddle method, in conjunction with the future crop pattern distribution.

The crop distribution is expected to alter in the next years. To estimate the future distribution of crops, a recent study of Ministry of Development (Development of Hydroinformatic systems and tools for water resources management, 2008) is taking into account the forthcoming changes to subsidy regimes imposed by the Common Agricultural Policy (CAP) from 2013. Taking also into account the increase of temperature, and consequently the increase of crop evapotranspiration, according to climate scenario A1B, the IWR are estimated for the period 2020–2030 to 126.1 million m^3 per year, named *Scenario 0 (100 %).*

The decrease of Irrigation Water Requirements (IWR) could be achieved by replacing the applied irrigation techniques from sprinkler irrigation to drip irrigation and by adopting a different crop distribution in the area including fallow fields or non-irrigated agriculture. Sprinkler irrigation is the dominant method in Asopos

RB while drip irrigation is applied only to 10 % of irrigated land. Considering that the efficiency of sprinkler irrigation varies from 0.6 to 0.8 and the efficiency of drip irrigation varies from 0.85 to 0.95, it follows that applying drip irrigation could save more than 20 % of irrigation water (0.7/0.9).

In order to decrease the above estimation of IWR in Asopos RB, the following two scenarios are considered:

Scenario 1 (75 %), IWR = 94.2 Million m^3

Applying drip irrigation to 75 % of irrigated land.
Assuming that 15 % of annual crops should be replaced by non-irrigated agriculture (e.g. wheat). Alternative, a fallow agriculture scheme should be adopted.

Scenario 2 (50 %), IWR = 62.6 Million m^3

Applying drip irrigation to 90 % of irrigated land.
Assuming that 55 % of annual crops should be replaced by non-irrigated agriculture (e.g. wheat). Alternative, a fallow agriculture scheme should be adopted.

7.2 *Impact of the Pressures on Water Resources and Proposed Measures for Their Management*

Reduced rainfall and higher temperatures will have an effect on water resources in the Asopos RB and on the wetlands functions performed along its channel, at the river mouth and in the Oropos lagoon and consequently on the values deriving from them. More specifically, future observations are expected to include:

- Increased water evaporation and plant evapotranspiration.
- Decreased recharge and renewal in the groundwater table due to decreased rainfall and increased evapotranspiration.
- Increased salinity in the coastal groundwater level.
- A more concentrated pollution load in the coastal aquatic ecosystem of the Asopos due to reduced dilution, with a negative impact on the nutrient transformation and removal function due to the resulting lack of water.
- Reduced discharge at the river mouth, with consequential effects on its biota; this, coupled with the increases in the concentration of pollutants, constitutes a serious hazard for many species of plants and animals in the riverine ecosystem.
- The food web support function and consequently the biological value of the region will be negatively affected. Climate change is among the main direct causes of loss of biodiversity and will have significant implications for the several constituents of biological diversity, namely ecosystems, species and the genetic diversity of species.

Moreover, climate change will lead to a reduction in the availability of water for irrigation, which will be due on the one hand to decreased rainfall and on the other to the lengthening of the dry period and increased demand for water for agricultural purposes. The net result will be that the region's water resources will be insufficient to meet the future water needs either of its ecosystems or of water users (urban water supply, irrigation, industry, livestock).

The measures proposed include:

- Reducing the amount of irrigated land and using crops requiring less water.
- Informing farmers about the rational use and management of irrigation water.
- Using recycled water in agriculture.
- Modernizing irrigation systems.
- Creating reservoirs.

Implementing these measures will reduce the amount of water required for irrigation, which in turn will lessen the impact of climate change on the availability of water resources in the region and consequently on the ability to meet the future water needs of the ecosystems and water users in the river basin.

Applying climate scenario A1B (2020/2021 to 2030/2031) to three irrigation needs scenarios (100 %, 75 %, 50 %) suggests that cutting irrigation requirements by 25 % will significantly slow the rate of fall in the groundwater table, although the rate of fall in the groundwater table is noticeably lower compared to the rate of fall observed when irrigation needs are met in their entirety. When irrigation needs are further reduced, to 50 %, the water table drops during the irrigation season but gradually recovers afterwards, and by the end of the simulation period is restored to roughly its initial level (Fig. 9.12).

Surface runoff in the Asopos basin increases by 9.5 % and 23.7 % respectively when irrigation requirements are reduced by 25 % and 50 % (Fig. 9.13). The increase in surface runoff resulting from the reduction of irrigation is particularly important during the irrigation season. When irrigation requirements are reduced by 25 %, surface runoff in the period from May to September increases by 17.8 %, while cutting irrigation requirements in half increases surface runoff by 42.8 %.

The fall in the groundwater table is also reduced and surface runoff increased when the three irrigation needs scenarios are run for the period 2070/2071 to 2080/2081. In this case, however, the drop in the groundwater table does not stabilize even when irrigation is cut in half (Fig. 9.14). As for surface runoff (Fig. 9.15), reducing irrigation to 75 % and 50 % of present levels results in an average annual increase of 9.7 % and 23.2 % respectively, and an increase during the irrigation season (May to September) of 22.9 % and 54.2 % respectively.

Furthermore, there are different economic tools that can assure the effective and equitable distribution of the water resources across competitive uses. These instruments can also contribute to internalize the externalities related to the use of a public good achieving a socially optimum result. These economic measures are presented in the Table 9.6 along with their main advantages and disadvantages.

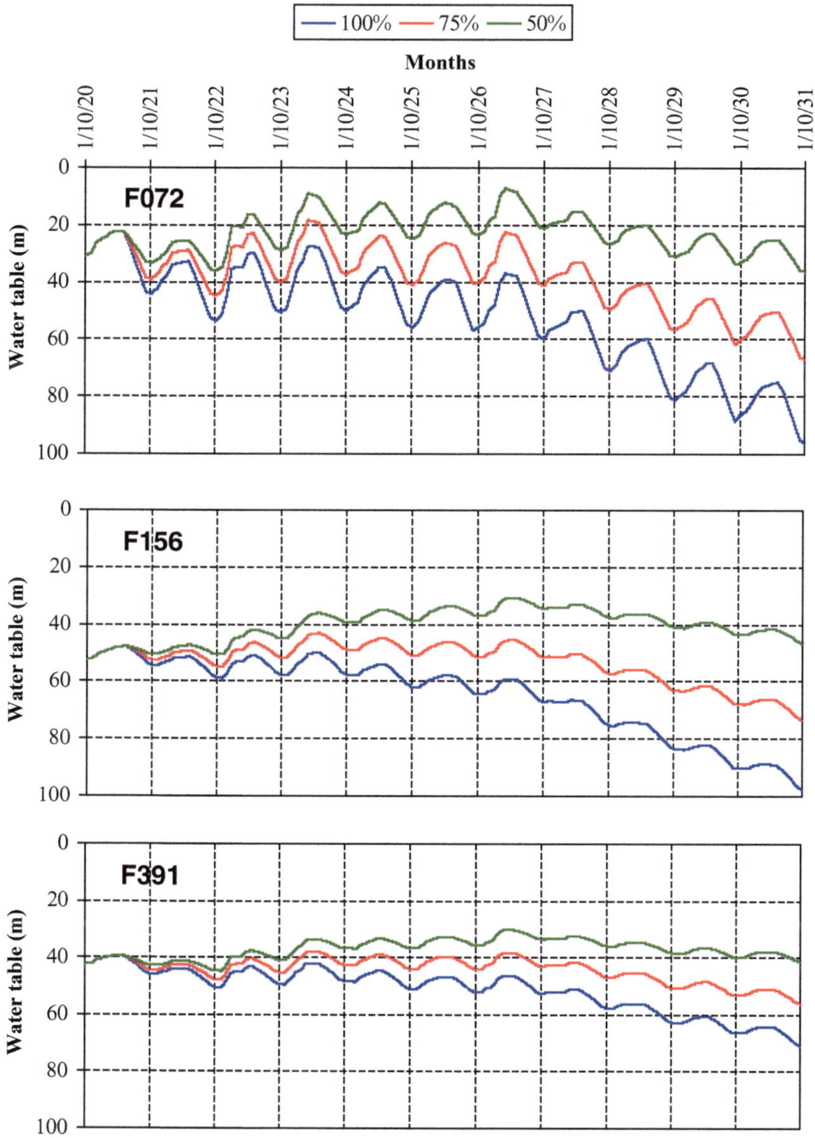

Fig. 9.12 Water table variation in Asopos RB under the climate scenario A1B (2020–2031) and the three irrigation requirement scenarios (100 %, 75 %, 50 %)

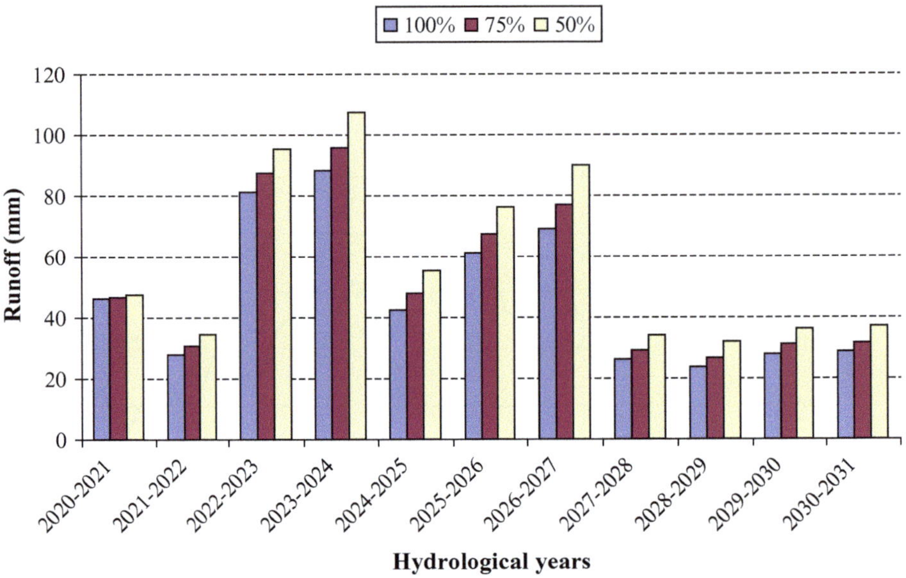

Fig. 9.13 Annual surface runoff in Asopos RB under the climate scenario A1B (2020–2031) and the three irrigation requirement scenarios (100 %, 75 %, 50 %)

Hence the search for the least cost set of measures may include: (i) economic instruments (e.g. abstraction/pollution taxes, tradable permits, subsidies), (ii) measures to increase awareness regarding water scarcity, aiming at reducing abstraction/pollution, (iii) direct controls on pollution dischargers, (iv) agri-environment programs providing financial and technical assistance for, e.g. reallocation of crop production mix over agricultural land, (v) adoption of water-saving technologies coupled with land-allocation restrictions, etc. and (vi) green investments e.g. pollution control and remediation, resource conservation and management, land use and infrastructure, renewable energy sources.

In order to restore wetlands functions and values in the region of the Asopos River and the Oropos lagoon, the measures for the management of water quantity in the Asopos basin will need to be supplemented by measures and interventions addressing water quality and control of point and non-point sources of pollution and taking into account existing socio-economic conditions in the region.

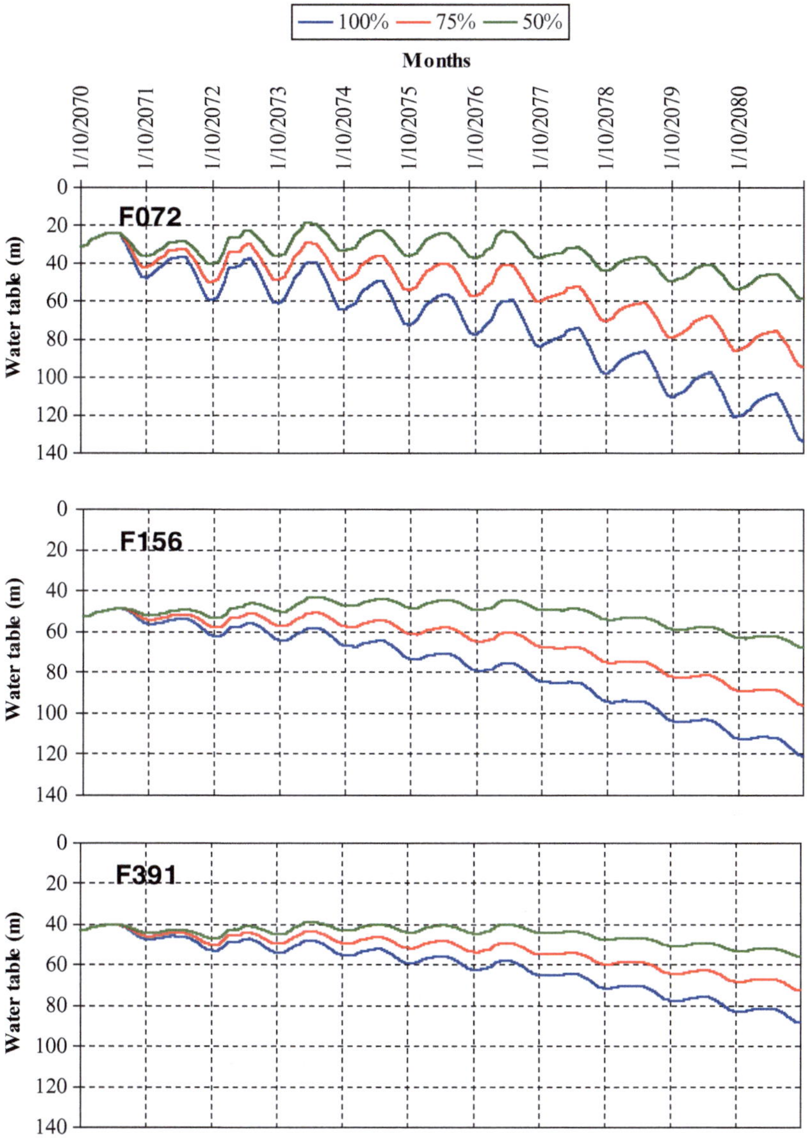

Fig. 9.14 Water table variation in Asopos RB under the climate scenario A1B (2070–2081) and the three irrigation requirement scenarios (100 %, 75 %, 50 %)

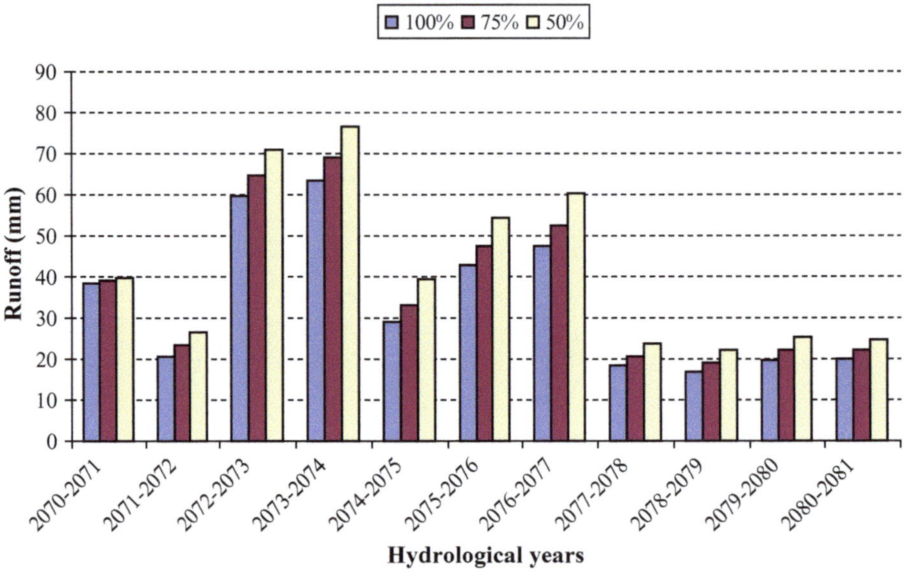

Fig. 9.15 Annual surface runoff in Asopos RB under the climate scenario A1B (2070–2081) and the three irrigation requirement scenarios (100 %, 75 %, 50 %)

Table 9.6 Classification of economic instruments[a]

	Advantages	Disadvantages
1. Standards and Quotas		Not economically efficient
2. Water abstraction charges	Adjustment of price signals to reflect actual resource costs; encourage new technologies; flexibility; generation of revenues	Low charges will have minimal impact on user behavior and will continue in resource overutilization
3. Pollution charges	Same as water abstraction charges; polluter-pays principle	Same as water abstraction charges
4. Subsidies on water saving measures	Readily acceptable	Financial constraints
5. Tradable permits	Quantity based targets that are able to attain least-cost outcome. Allows flexibility.	May entail high transaction costs
6. Voluntary agreements	Readily acceptable	Needs high environmental awareness
7. Liability legislation	Assess and recover damages ex-post but can also act as prevention incentives	Require an advanced legal system; high control costs; burden of proof

[a]Report on the implementation of Article 5 of the WFD (2008)

References

Adamus, P. R., Clarain E. J., Jr., Smith, R.D., & Young R. E. (1987). *Wetland valuation Technique (WET), Volume II: Methodology. US Army Corps of Engineers, Waterways Experiment Station, Vicksburg, Mississippi. Operational Draft Technical Report Y-87 and Federal Highway Administration* (FHWA-IP-88-029).

Bakos, A. (2009). *Qualitative analysis of surface and groundwater in the hydrological basin of Asopos river supported by the detection of carcinogenic substance hexavalent chromium.* Thesis, Department of Natural Resources Management & Agricultural Engineering, Agricultural University of Athens, 217p. & Annexes.

Brinson, M. M. (1993a). *A hydrogeomorphic classification for wetlands.* Wetlands Research program Technical Report WRP-DE-4. US Army Corps of Engineers Waterways Experiment Station, Vicksburg.

Brinson, M. M. (1993b). Changes in the functioning of wetlands along environmental gradients. *Wetlands, 13*, 65–74.

Crawford, N. H., & Linsley R. K. (1966). *Digital simulation in hydrology: The Stanford Watershed Simulation Model IV* (Technical Report No. 39). Department of Civil Engineering, Stanford University, Stanford, CA, 210p.

DHI. (2009). *MIKE SHE user manual: A dynamic modelling system for integrated groundwater and surface water resources*. Denmark: Danish Hydraulic Institute.

Economou, A., et al. (2001). *Study of fisheries management in lakes (natural and artificial), utilization of water resources in mountainous and less favoured areas in Prefectures of Aitoloakarnania, Evritania, Karditsa, Viotia, Arkadia, Elia and Achaia: Phase A'*: Final report, 599p.

EMEKA-Climate Change Impacts Study Committee. (2011). *The environmental, economic and social impacts of climate change in Greece.* Athens: Bank of Greece-Economic Research Department. 520p.

Frangopoulos, I., Alexiadou, M., & Panagopoulos A. (1992). *Definite hydrogeological survey in the district of Thiva.* Thebes: Ministry of Rural Development and Food.

Giannoulopoulos, P. (2008). *Hydrogeological – hydrochemical survey on groundwater quality in Asopos basin, Prefecture of Viotia.* Athens: Institute of Geology and Mineral Exploration, 74p.

Kapsomenakis, I., Douvis, K., Giannakopoulos, X., Zanis, P., Tselioudis, G., Repapis, C. C., & Zerefos C. S. (2011). *Scenarios of anthropogenic intervention in climate change and the Prudence and Ensembles projects.* Athens: Bank of Greece – Climate Change Impacts Study Committee.

Katsavouni, S., Seferlis, M., & Papadimos, D. (2012). *Assessment of functions and values of Asopos wetland* (2nd ed.). The Goulandris Natural History Museum – Greek Biotope/Wetland Centre. Thermi, Greece. 79p (In Greek).

Maltby, Ed. (Eds.). (2009). *Functional assessment of Wetlands. Towards evaluation of ecosystem services.* Woodhead Publishing Ltd, Granta Park Great Abington, Cambridge, UK.

Marble, L. M. (1992). *A guide to wetland functional design.* Roca Raton/Ann Arbor/London: Lewis Publishers.

Nakićenović, N., Alcamo, J., Davis, G., de Vries, B., Fenhann, J., Gaffin, S., Gregory, K., Grubler, A., Jung, T. Y., Kram, T., La Rovere, E. L., Michaelis, L., Mori, S., Morita, T., Pepper, W., Pitcher, H., Price, L., Raihi, K., Roehrl, A., Rogner, H.-H., Sankovski, A., Schlesinger, M., Shukla, P., Smith, S., Swart, R., van Rooijen, S., Victor, N., & Dadi, Z. (2000). *IPCC special report on emissions scenarios*. Cambridge, UK: Cambridge University Press.

Report on the implementation of Article 5 of the WFD, Hellenic Ministry of the Environment, Physical Planning and Public Works, Athens March 2008, Prepared by RESEES – Research on Socio-Economic and Environmental Sustainability – Team. Available at http://www.aueb.gr/users/koundouri/resees/en/aswposprojen.html (In Greek).

Trewartha, G. T., & Horn, L. H. (1980). *An introduction to climate* (5th ed.). New York: McGraw-Hill.

Tsarabaris, C. (2010). *Hydrogeological regime in the upper part of Asopos River. Investigation of qualitative aspects of groundwater degradation*. Thesis, Agricultural University of Athens, 147p.

Tzali, M., Fric, J., & Promponas, N. (2009). *The birds of wetlands in Attiki. Bird Monitoring Programme on Wetlands of Attiki*. Athens: Hellenic Ornithological Society. 56p.

Yan, J. J., & Smith, K. R. (1994). Simulation of integrated surface water and groundwater systems – Model formulation. *Water Resources Bulletin, 30*(5), 1–12.

Chapter 10
Creating the Institutional Background to Support the Implementation of the Policy Manual

P. Koundouri, N. Papandreou, M. Stithou, and O.G. Dávila

Abstract Following Water Framework Directive's (WFD) implementation a national water policy that provides the institutional background to achieve the goals of the suggested policy is required. This Chapter emphasizes on the issues of water policy that should be addressed and provides reform options that could lead to the sustainable use of water resources. Efficient use of water is an important issue in Europe; however poor governance seems to hinder effectiveness of the designed policies. After a review of the Greek experience with water reforms, the "state oriented" regime in Greece is presented along with the legislative frame and institutions that it entails. Then institutional improvements are discussed. In particular, the institutional reforms move away from water policies that favor fragmentation and lack of coordination among the involved actors in the decision making process. These reforms also embrace and promote the growing concerns on environmental aspects and "sustainable" water management suggesting integrated approaches. In the suggested framework of action the transfer of competencies from the state to the utility and to other actors, induced by the WFD, makes more imperative the existence and enforcement of a legislative and administrative framework that could accelerate the co-operation among the involved parties. The chapter closes with commenting on how the case of Asopos is related to the other Greek basins that are at risk of failing to meet WFD's obligations.

Keywords Water Framework Directive • Water policy • Sustainable water management • Water legislative and administrative framework • Institutional reforms

P. Koundouri (✉) • M. Stithou • O. Dávila
Athens University of Economics and Business, Athens, Greece
e-mail: pkoundouri@aueb.gr; stithoumavra@yahoo.com; osiel.davila@soas.ac.uk

N. Papandreou
Andreas G. Papandreou Foundation, Athens, Greece
e-mail: tulayev@gmail.com

P. Koundouri and N.A. Papandreou (eds.), *Water Resources Management Sustaining Socio-Economic Welfare*, Global Issues in Water Policy 7,
DOI 10.1007/978-94-007-7636-4_10,

1 Introduction

As it was also highlighted in Chaps. 3 and 4 in Greece the provision of water supply and sewerage is treated as a public service. Greece updated its water management framework with the 3199/03 Law. By the 1980s, the policies adopted by Greece on water resource protection mainly dealt with aspects of public health protection, while before the 3199/03 Law was put into force the legislative framework of the country included Law 1739/1987 on Water Resources Management and the Environmental Protection Law 1650/1986. The 1987 Law established the institutional framework for the management of water resources in Greece and it was the first law defining principles for more intersectoral and coherent water management. In particular, water scarcity was recognized as an issue of national significance, public domain was enlarged with the introductions of water permits for water abstractions and for water works putting water management on the hands of the State and water use under public control. Other changes that this Law brought were that it promoted water policy planning by adopting a river basin management approach, some stakeholder participation as well as the protection of the aquatic environment from water over-use (Kampa 2007).

However, despite the innovative and integrated approach its complexity made its full implementation difficult. In addition, water quantity and water quality issues were treated separately by different laws and authorities (Kampa 2007), while the law was also criticized for not including adequately the principle of water resource protection from an environmental perspective (Terzis 1997). Thus, the national water regime in this phase remained in reality complex and fragmented and undermined by a non-supportive institutional context. As highlighted in Kampa (2007):

> The centralization of the Greek politico-administrative system combined with the lack of resources of the administration and the top-down style of the integration attempt hindered the set-up of a coordinated balanced network of central and decentralized water management authorities. Integration was also hindered by integration unfavorable dominant values and perceptions in the water sector, such as the absence of cooperative water policy style and intersectoral planning, the fragmented understanding of water problems by different sectors and the low societal awareness of the environmental dimension of water resources. (p. 115)

As a result in order for the measures presented in Chap. 9 and other previous chapters to be effective, the above policy pitfalls of water governance model need to be addressed. The following section presents the current legislative frame and involved institutions.

2 The legislative Frame and Institutions

As far as the Greek legislation for the management and protection of water resources is concerned, it is noted that it is mainly based on the European legislation since until 2000 there was not a well-defined integrated water resources policy.

According to the Law 3199 of 2003 which incorporates the WFD (CEC 2000) the need for the National Council of Water Resources has emerged in order to implement a national water policy. The recently established (in 2010) Council is responsible for the planning of policy for the management and protection of country's water resources. It also monitors and controls the implementation of the planned water policy and approves the national programs for the protection and management of the water resources of the country.

Furthermore, the National Water Committee under the suggestions of the Minister of Environment, Energy and Climate Change, is responsible for the formulation of the government's policy and also for the submission of an Annual Report to the European Parliament which includes the country's aquatic environment status, the implementation of legislation for the water protection and management and its conventionality with the European Framework.

Within the framework of the voted by the Greek Parliament Law (3199/2003) a Special Secretariat within the Ministry of Environment Energy and Climate Change was established. "The Special Secretariat for Water is responsible for the development and implementation of all programs related to the protection and management of the water resources of Greece and the coordination of all competent authorities dealing with the aquatic environment. The implementation of the Water Framework and the Marine Strategy Directives as well of the related daughter Directives fall within the scope of the activities of the Secretariat. The Secretariat, in collaboration with the Regional Water Authorities, formulates and, upon approval by the National Council for Water, implements the River Basin Management Plans and the national monitoring program. The Secretariat is composed of four Directorates and is headed by a Special Secretary, appointed by the Ministry of Environment, Energy and Climate Change and the Government."[1]

The following figure presents the authorities and organizations relevant to water supply, management, control and authorization in Greece (Fig. 10.1).

The Central Water Directorate acts on a technical level and in particular it[2]:

1. Develops national programs to protect and manage the water resources and inspects their implementation;
2. Coordinates Services and National Bodies and participates in Competence Community Institutes, on water management and protection issues;
3. Recommends water pricing policies
4. Recommends legislative and administrative measures for water management and protection
5. Observes water quality and quantity in national level
6. Manages meteorological and hydrological database and information
7. Observes all Regional Water Directorates

[1] http://www.ypeka.gr/Default.aspx?tabid=246&language=el-GR

[2] http://www.waterincore.eu/deliverables/03_02_01_en.pdf

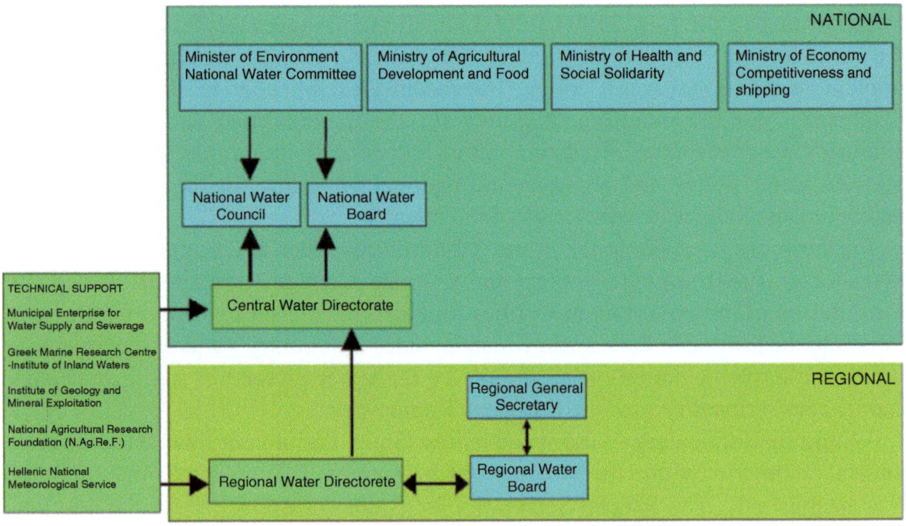

Fig. 10.1 Administrative schemes-bodies of water management in Greece (Source: MED Programme 2007–2013, http://webcache.googleusercontent.com/search?q=cache:VNq9HQERq9QJ:www.waterincore.eu/deliverables/03_02_01_en.pdf+&cd=7&hl=en&ct=clnk&gl=uk)

8. Prepares a detailed report of river basins characteristics, impacts of human activities, economic analysis of water use
9. Sets up protected areas national register

At the regional level the Regional Water Directorate specializes and applies long- and intermediate-term protection and management programs of rivers basins, evaluates Management Plans and Measure Programs for the water basins. Also is in charge of formatting register of protected areas and cares for the substantial participation of the public. It implements Monitoring programs of the qualitative and quantitative parameters of the water and applies programs of measures on protection from pollution and restoration of water and specializes in the national pricing policy and political costing of the water usages.

The Regional Water Board gives consultative response before the approval of the Management Plan and expresses its opinion on every water protection and management topic that is submitted by the General Secretary of the Region and finally Publishes Management Plans, in order the public to be informed on its content and participate in the public consultation.

All other institutes (Greek Marine Research Centre-Institute of Inland Waters Greek Biotope/Wetland Centre, Hellenic National Meteorological Service Institute of Geology and Mineral Exploitation, National Agricultural Research Foundation) are related with research activities and collaborate in water management with the national and regional bodies.

At local level the Regional Units of the new administrative divisions of Regions (Kallikratis Law 3852/2010), have an Authorization, Inspection and Surveillance role. Municipal Enterprises for Water Supply and Sewerage is the competent authorities responsible for the control, pricing and management of water supply, sewerage and wastewater treatment. An important parameter for the right implementation of the WFD is related to the coordination and smooth operation of the various authorities. In order to face these weaknesses the Kallikratis Law 3852/2010 attempts to make more flexible the whole framework of the new decentralized governance. In this context the State Administration is responsible for the design of the strategic plan of protection and management, while the elected regions need to implement it.

The Responsible Bodies for the irrigation network management are the General Organizations of Land Reclamation (GOLR) which operate under the inspection and surveillance of the government, Ministry of Rural Development and Food. GOLR inspects, coordinates and give guidance to Local Organizations of Land Reclamation (LOLR), which are responsible for local irrigation networks management.

3 The Institutional Background of a Reformed National Water Policy

According to Saleth and Dinar (2005), water institutions can be defined as "rules that together describe action situations, delineate action sets, provide incentives and determine outcomes both in individual and collective decisions related to water development, allocation, use and management." In this context, water institutions are considered deeply rooted in the cultural, social, economic and political circumstance of each nation. In consequence, institutional change can only arise from such circumstantial factors, the individuals that make up the institutions and their attitude towards them. Also important for institutional change are the estimations of transaction costs: real and monetary costs that are provoked by changes in the regulatory, monitoring and enforcement mechanisms related to water allocation and management. Scale economies are also relevant because transaction costs are subject to scale economy effects from institutional linkages and increasing returns in water institutional performance. The institutional transaction cost theory can be used to explain the endogenous (internal to the water sector) and exogenous factors that affect the institutional change process. Hence, it is important when analyzing the institutional background of a reformed national water policy to take into consideration all the previously mentioned factors (the socio-economic and political context as well as related transaction costs) since they have a huge impact on the way the institutional change in the water sector is justified. The authors argue that "political economy theories and transaction cost approaches are useful to indicate the rationale for reforms." Nevertheless, from the perspective of designing and implementing purposive reform policies, the stage-based perspective of the change process is considered to be more promising in view of its insights into how and when to intervene within the reform change process. In a previous paper, Saleth and Dinar (2004) developed the stage-based perspective of institutional change. It provides a more complete description of the institutional change process. They identify four main stages of institutional change: (1) Mind change, (2) Political articulation, (3) Institutional change and (4) Actual impact. These stages progress as a circular process. They point that "in different stages, the change process is mediated by mechanisms such as instrumental (or reference point-based) subjective evaluation, information flow and learning externalities, political lobbying and bargaining, organizational power and politics, and behavioural changes and performance expectations."

Greece opted so far for the social planning or public type of water allocation. Furthermore, the lack of well-structured state water authorities to supervise water related activities (e.g., irrigation) of private users and a gap in the regulation of the access to water resources has led to improper water resources management. It is argued that a "state oriented" regime apart from other drawbacks it usually fails to create incentives for users to conserve water and improve efficiency (Dinar et al. 1997). Therefore it is necessary for Greece to move as soon as possible towards a water pricing reform in order to improve water management and comply with European legislation.

However, as stated in Kampas (2007) improvements are needed in the institutional elements of the Greek national water regime. These improvements include primarily more coherence in the system of Greek public governance, meaning coordination between administrative levels at river basin scale, establishment of public participation processes in water planning, coordination of perspectives of multiple actors and sectoral authorities and concentration of water competence to fewer authorities. Secondly, issuing of water abstraction permits should be linked to an intersectoral water planning system and there should be better control of water abstractions. Nevertheless, it is acknowledged that implementation deficiencies need to be addressed in order to achieve this aim which is not particularly easy especially when facing severe economic crisis and tight budgets. Finally, it is argued that there is a need to better coordinate groundwater ownership rights and groundwater use rights and eliminate private ownership of groundwater. In addition, coordination between public policy and property rights could be enhanced by investing in enforcement mechanisms that aim to oversee policy-based rights to abstract or pollute water.

As also emphasized by Mimikou (2005) the established polyarchy disallows an integrated approach that would be necessary to plan and implement a national water policy. The cost of implementing the necessary measures as dictated by the WFD in such a context with many involved responsible institutions would be substantial. At this critical point a rational reorganization of the entire water sector in Greece is deemed necessary and the author suggests the following actions as the most important (Mimikou 2005, p. 321):

- Apart from the creation of a central institution such as that of the National Council of Water Resources a National Water Research Institute should support technologically and scientifically the former following the existing organizational schemes of other European Countries.

- The application of the WFD should immediately be stimulated in a high scientific and technical level, along with a proportional activation and representation in Europe – in all the sectors offended by the Directive.
- The water uses should be regarded with a more balanced and rational sharing of the available resources. For the rural use, in particular, a rural national water policy should be immediately planned, at least concerning the proper choice for cultivation of land based on the region's water availability, the observance of the laws regarding exploitation and pollution of underground waters, the vital reduction of water losses and water consumption for irrigations (amendment of irrigation practices etc.) and the briefing and guidance of citizens and farmers on these subjects.

4 The Asopos Paradigm and Common Areas of Intervention Across Other Greek Basins

The Asopos River catchment basin is an area among those with the highest industrial activity in Greece and at the same time one of the most polluted areas in the country. The unregulated and uncontrolled operation of businesses in the informal industrial area of Oinofyta – Schimatari for years as well as several shortages in infrastructure have contributed so that the aquifer of the area to be contaminated by heavy metals, including hexavalent chromium. The agricultural sector has also responsibility for the current situation as it is linked to nitrates pollution and water abstraction although it also suffers the economic consequences of water quality degradation through increased cost and reduced demand for the area's products. Finally, the residents of the basin face health issues and downgraded quality of life.

It should be noted that Asopos differs to other river basins and The Ministry of Environment Energy and Climate Change (YPEKA) has declared the issue "Asopos" as a case requiring "Special National Intervention". YPEKA publicly acknowledged the environmental damage occurred in the greater area of Asopos River (Oinofyta – Schimatari) and its consequences on Public Health and tackling this problem by taking serious measures has been considered as key priority by the Ministry's political leadership.

As presented in Table 10.1 through the chapters of this book starting from evaluating the profile of the area with regard to water use by sectors, different methods (choice and laboratory experiments, benefit transfer and estimation of distance functions) were presented and the results of their empirical application enabled us to estimate the environmental and social dimensions of the problem and importantly inform water policy along the lines of WFD.

Regarding agriculture, as shown in Chap. 3 considering that the average price for water consumption for irrigation is €13.73 per stremma of irrigated land (Koundouri et al. 2009) the value of water related to its use component for irrigation is estimated at €755,850. In addition, as presented in Chap. 5 the estimated shadow price of in situ groundwater is €0.154 per m^3 of water. Considering the fact that the demand for irrigation from groundwater is estimated at 51,565,788 m^3 per year (YPEKA 2012)

Table 10.1 Cost recovery in Asopos river basin (RB)

Users	Value of water	Polluter pays principle		User pays principle
Industrial	€439,385	€2,690,000[a]	€882,200 to €1,649,480[b]	
Agricultural	€755.850			€818,799
Domestic	€758,980			
Total	€1,954,215			

[a]Benefit transfer method
[b]Choice experiment method

the resource cost associated with this particular use is considerably high (€8M). Furthermore, taking also into consideration that a part of the estimated environmental cost (€882,200) is a result of as well agricultural pollution it makes the reform in the pricing system for irrigation water even more obvious in order to mitigate the negative impacts of the existing system.

Following Latinopoulos (2005) under the current status private users of irrigation water bear their own capital and operational costs, which can be up to a maximum of €0.25–0.40 per m^3. On the other hand, users of public collective projects pay a usually a flat rate per hectare of irrigated land to cover the administrative as well as the maintenance and operational costs of the projects. In average this fee ranges from €120 to €500 per ha, which is roughly equivalent to €0.02–0.08 per m^3. In the case of Asopos considering a total irrigated land of 55,051 str. (5,505 ha) the value of irrigated water could range from €660,612 to €2,752,500. As capital costs of irrigation water projects are covered totally by state funds, while external costs are usually not taken into account, the low cost recovery of public water supply combined with the high opportunity cost of irrigation water portrays a very inefficient and non-equitable system of agricultural water provision. Agricultural water prices neither cover the cost of supply nor provide sufficient conservation incentives (OECD 2009).

Following the analysis of Chap. 4 and considering a modest total of 23,000 residences which are permanently connected with the public system of water supply, the average fixed fee of water supply of €30.26 per year, the average charge of the 100th m^3 of €0.60 per m^3, and a 10,494,930 m^3 annual water consumption, the value of water for domestic use is about €758,980. Finally, regarding the value of water for industrial use as reported in Chapter 3 (Table 3.7) it is calculated at €439,385 considering the areas of Schimatari and Oinofita.

As far as the environmental cost is concerned according to the table it ranges from €2,690,000 estimated by the Benefit Transfer method to €882,200 and €1,649,480 estimated by the Choice Experiment method and depending on the number of households used for aggregation. The total number of dwellings in Asopos basin reaches 43,223 residences from which at least 43,000 residences are considered to be connected to or going to be connected to the water supply network. In addition, the 23,000 regular residences are permanent residences and therefore it is regarded that are permanently connected with the network of water supply and consume water during the whole year. The first aggregation is calculated based on the 23,000 households while the second on 43,000. It is also reminded that these estimates consider only the ecological damage and not the economic and social.

The applied methods give an indication of how society values environmental improvements. By looking at the difference between the value actually paid to the supplier for water services and the ecological damage as well as the opportunity cost of use it becomes obvious that the cost recovery by applying the polluter pays principle and the user pays principle can internalize the externalities and create incentives for efficient use of the resources. Hence, the charges should reflect the real cost of water and they should not only incorporate the cost of abstraction, transfer and maintain of the network but also the environmental and resource cost. However, that does not necessarily lead to higher charges in all the sectors. The water pricing of competent bodies can and ought to be socially equal. This can be potentially achieved through the gradual charges, reductions for the vulnerable social groups and certainly as discussed through the implementation of the polluter pays principle so as the real cost of water not to be borne unequally and without discriminations by whole of society. A river basin management plan for Asopos could include apart from the spatial planning of the industrial areas of Oinofyta and Schimatari and the establishment of waste management infrastructure the adoption of economic instruments as presented in Chap. 9. However, it is important to become common knowledge that the immediate adoption of measures to prevent and remediate the environmental damage will entail costs which will now be borne by the polluters. That is the cost for the creation of a particular infrastructure will be borne by the producers of hazardous waste, in application of the principle of "whoever pollutes, pays". At this point it is reminded that although results represent the Asopos reality, improvements in cost recovery mechanism and estimation of its elements is relevant for other river basins as presented in Chap. 1. Other necessary preconditions are the establishment of clear rules, guidelines and procedures, upon which pollution sources will be monitored and industries will be bound to operate. Furthermore, an inventory of quantities and types of hazardous waste would enhance management considerably.

Another important issue that needs to be addressed is that the shared competence between multiple services at central and regional level in relation to Asopos River and the greater area, as well as between multiple Ministries (YPEKA, Ministry of Health, Ministry of Rural Development and Food, Ministry of Infrastructure, Transport and Networks) creates difficulties in addressing the problems immediately and effectively and causes delays in the implementation of measures that fall under the shared competence of these bodies. It is needless to point out that this is an issue relevant for the rest of river basins. In the case of Asopos due to its special characteristics (industrial pollution with hexavalent chromium and other dangerous for the environment and human health heavy metals), the adjustment of the water policy of the competent bodies has to be applied in a common and single base. Therefore, in order to achieve a full cost recovery in the whole Asopos area, the competent bodies of water provision should adopt a single water policy characterized by a common escalating charge and fixed costs.

The establishment of Regional Committees at decentralized administration level, promoted by the Coordination Office for the Implementation of Environmental Liability, the priority of which will be the prevention and remediation of environmental damage caused in Asopos River area, in close cooperation with the Decentralized Administration of Thessaly – Sterea Ellada and the competent Regional Committee for the Management of Environmental Damage, will make easier to identify several environmentally liable enterprises.

Overall, the Asopos case represents a reality that other river basins face with similar or of less magnitude problems. In order to improve the knowledge regarding the degree of cost recovery in each hydrometric area in Greece and the country's capability for the full implementation of the European legislation in the future, the submitted report on the implementation of Article 5 of the WFD (2008) suggests the following policy interventions:

- To adopt common water charges for the total of the DEYA[3] (water supply) and TOEB[4] (irrigation)
- To allocate responsibilities to the competent bodies regarding issues of management and protection of water resources in order to avoid overlapping of responsibilities
- To create a data base with quantitative and qualitative parameters for monitoring the change over time as well as the creation of a computerized registration system of the necessary information for performing the economic analysis
- To evaluate the various tools of economic policy (taxes, quantitative restrictions, subsidies) as to their ability to provide incentives for efficient management of water resources at the lowest possible cost.

It becomes evident that in Greece the need for sustainable management of water resources is deemed necessary and immediate action is needed due to the pollution problems, the increased demand for water and the challenges that climate change will bring. It could be argued that one of the reasons for the excessive use of water resources is the fact that water pricing does not reflect the real cost of the good. Dinar (2000) explains that among the most important measures developed to encourage the efficient use of hydric resources is water pricing reforms. The right pricing, following the WFD requirements, will lead to the improvement of water quality and to its proper use according to the needs of demand. Nevertheless, it is important to keep in mind that the design and implementation of pricing reforms is a very complex process that is affected by several variables that may prove difficult to define and model. For example, there are several examples in the literature that indicate that political pressure could have an impact on the implementation of water pricing reforms and thus the result may diverge from the original objectives. Therefore, it should be noted that the property regime has to be taken into account when the reform of water pricing is conducted and should include water consumers, suppliers and regulators. In order to the reform to be successful, current

[3] Municipal Enterprises for Water Supply and Sewerage.

[4] Regional Organizations of Irrigation Networks.

bureaucracies and all the relevant groups of interests and stakeholders should be involved in the design and implementation of all the pricing reform stages. At the same time, a targeted and accurate dissemination of information is vital during the reform process.

An eminent reality for Greece is the passage to the era of the management of water demand in each one of its river basin districts, as the country needs to face the unequal distribution and availability of resources through management in each district rather than the transfer of water from the surplus to deficient districts or the creation of structural interventions (e.g., dams, river diversions, deep drilling etc.). The construction of desalination plants and/or the reuse of water through the promotion of renewable resources will be beneficial especially in areas with water deficit. The right management of the wastewaters coming from the industries and from the network of sewage is also of concern as well as the replacement of irrigation methods with more environmentally friendly practices and the replacement of water consuming crops in arid areas.

Furthermore, Greece should promote more the public consultation which is deemed very important for the finalization of the management plans of the river basins. It is also worth noting that the promotion of awareness of the ecological management of water should be a primary concern.

5 Conclusions

The main characteristic of water supply in Greece is the existence of a "state oriented" regime of fragmented functions combined with the lack of an efficient and equitable water pricing system. Greece transposed the EU WFD into domestic law in 2003 and in order to implement it, it created 13 Regional Water Directorates and a specialized Central Water Agency. This governmental authority is in charge of the definition and oversight of the national water policy.

With regard to water management according to OECD (2009) report the challenges that the country faces are excessive pumping of groundwater, illegal abstractions and discharges, difficulty in enforcing the regulations and water permit conditions, water losses in urban and irrigation water conveyance systems, agricultural water prices that neither cover the cost of supply nor provide sufficient conservation incentives, longstanding pollution hot spots and lack of implementation of a plan to control discharges of dangerous substances. Other aspects that need attention are the ecological aspects of water quality and improvements in the monitoring system.

Report's recommendations focus on the (i) continued effort to comply with EU WFD, (ii) the need to formulate and implement a national irrigation policy that promotes the rational use of water, the reduction of groundwater abstractions and the improvement of irrigation efficiency and practices in both communal and private irrigation networks, while ensuring that all water abstractions are properly licensed, (iii) further improvement of wastewater management (iv) intensified efforts to reduce water pollution by dangerous substances, to prevent illegal discharges of wastewater, and clean up pollution hot spots, (v) introduction of new measures to improve the allocation of water to ensure water flows to the highest-value uses and (vi) increase public awareness and understanding, particularly among farmers, of the economic, social and environmental aspects of water management. In addition, as noted in Latinopoulos (2005), properly designed water management policies should aim not only at environmental and economic goals but should also consider social impacts to the main stakeholders. Hence, in this framework, a water pricing structure should at the same time ensure equity at low price levels and a more strict and efficient administration that will safeguard the sustainability of water resources.

However, as it has been emphasized an important drawback in the achievement of the above targets is the fragmentation of functions across the authorities responsible for water management leading to duplication, making coordination a challenging task and generally leading to inefficiencies.

Although improvements have been noticed as highlighted in Kampa (2007) political effort is needed in terms of better enforcement of regulations and overcome of political inertia on different levels (national, regional, local), adoption of the polluter pays and full cost recovery principles and strengthening of the environmental awareness of Greek society. Furthermore, alternative less engineering-intensive management solutions should be gradually investigated and promoted and a nation-wide water information basis should be established. In addition, it is important that the level of coordination and co-governance of central and decentralized authorities in Greek water management to be strengthened. In this framework, actor cooperation, integration-supportive institutional interfaces and joint problem perceptions will enable Asopos and the rest of river basins in Greece to achieve their WFD targets.

However, it should be also highlighted that the institutional reforms associated with changes in the distribution of power and benefits can create political opposition and the larger the number of interest groups the more complicated the implementation of the above measures/reforms, as they have a role to play in both the design and implementation stages of a reform (Dinar 2000). In any case some of the factors that can help implement successful reforms include the commitment of a strong government, the use of alternative policy measures to allow for sustainable reform consequences, an efficient reform program leading to low transition costs, the implementation of a safety net for vulnerable social groups (Williamson and Haggard 1994), economic rationality, political sensitivity and constant attention to political-economic interactions and social-institutional factors (Cordova 1994). The timing of a reform is also important and may be for the case of Greece the public perception of the crisis (Williamson 1994) may create conditions under which it is politically possible to undertake the necessary reforms. As Saleth and Dinar (2005) argue “fiscal crisis provides a favorable context for implementing even a radical program with the least political opposition. Similarly, when the water sector reform forms part of larger political or economic reforms, its implementation becomes easier owing to synergic effects and scale economy benefits from the larger program.” (p. 8).

Overall, in Greece, there is an evident need for sustainable management of water resources. Because of pollution problems, water scarcity and higher demand of water immediate action is required. The fact that water pricing does not reflect the real cost of the good is one of the reasons for the excessive use of this resource. As it was explained in Sect. 4, among the most important measures developed to encourage the efficient use of hydric resources we find water pricing reforms. However, it is important to keep in mind that the design and implementation of pricing reforms is a very complex process. As it was discussed in Sect. 3, any institutional change is shaped by the socio-economic and political context of a region as well as related transaction costs related to the institutional change. Thus, the current property regime has to be taken into account when the reform of water pricing is conducted and should include water consumers, suppliers and regulators. In order to the reform to be successful, current bureaucracies and all the relevant groups of interests and stakeholders should be involved at all the reform stages. Of course, in the Greek case, the legislative frame and institutions described in Sect. 2 should be the starting point for any water pricing reform and institutional change in the water sector.

References

Commission of the European Communities (CEC). (2000). *Directive of the European Parliament and of the council establishing a framework for community action in the field of water policy*. 1997/0067 (COD), C5-0347/00.

Cordova, J. (1994). Mexico. In J. Williamson (Ed.), *The political economy of policy reform*. Washington, DC: Institute for International Economics.

Dinar, A. (2000). *The political economy of water pricing reforms. Published for the World Bank*. New York: Oxford University Press.

Dinar, A., Rosegrant, M. W., & Meinzen-Dick, R. (1997). *Water allocation mechanisms – Principles and examples*. Washington, DC: The World Bank.

Kampa, E. (2007). *Integrated institutional water regimes: Realisation in Greece*. PhD thesis, Technische Universität Berlin, Germany.

Koundouri, P., Remoundou, K., & Kountouris, Y. (2009). A *note on the implementation of the economics of the EU Water Framework Directive under data limitations: A rapid appraisal approach*. DEOS Working Papers, Athens University of Economics and Business, Athens.

Latinopoulos, P. (2005). Valuation and pricing of irrigation water: An analysis in Greek agricultural areas. *Global NEST Journal, 7*(3), 323–335.

Mimikou, M. A. (2005). Water resources in Greece: Present and future. *Global NEST Journal, 7*(3), 313–322.

OECD. (2009). *Environmental performance reviews*. Greece: OECD.

Report on the implementation of Article 5 of the WFD, *Hellenic Ministry of the Environment, Physical Planning and Public Works*, Athens March 2008, Prepared by RESEES – REsearch on Socio-Economic and Environmental Sustainability – team available at: http://www.aueb.gr/users/koundouri/resees/en/aswposprojen.html

Saleth, R. M., & Dinar, A. (2004). *The institutional economics of water: A cross-country analysis of institutions and performance*. Cheltenham: Edward Elgar.

Saleth, R. M., & Dinar, A. (2005). Water institutional reforms: theory and practice. *Water Policy, 7*, 1–19.

Terzis, C. (1997). *Expertise concerning the introduction of integrated water management structures in Greece*. Final report financed by the EC DG XVI B.2 through the Technical Assistance Management Company. Contract number 6807/1-3-96. Athens.

Williamson, J. (Ed.). (1994). *The political economy of policy reform*. Washington, DC: Institute for International Economics.

Williamson, J., & Haggard, S. (1994). The political conditions for economic reform. In J. Williamson (Ed.), *The political economy of policy reform*. Washington, DC: Institute for International Economics.

YPEKA (Ministry of Environment Energy and Climate Change). (2012). *Management Plan of the River Basins of the River Basin District of the Eastern Sterea Ellada: Analysis of anthropogenic pressures on the surface and groundwater water bodies*. January 2012. Available at: http://www.scribd.com/doc/97686793/P08-WD07 (In Greek).

Index

P. Koundouri and N.A. Papandreou (eds.), *Water Resources Management Sustaining Socio-Economic Welfare*, Global Issues in Water Policy 7,
DOI 10.1007/978-94-007-7636-4,

MIX
Papier aus verantwortungsvollen Quellen
Paper from responsible sources
FSC® C105338

If you have any concerns about our products,
you can contact us on
ProductSafety@springernature.com

In case Publisher is established outside the EU,
the EU authorized representative is:
Springer Nature Customer Service Center GmbH
Europaplatz 3, 69115 Heidelberg, Germany

Printed by Libri Plureos GmbH
in Hamburg, Germany